Ecology: A Very Short Introduction

VERY SHORT INTRODUCTIONS are for anyone wanting a stimulating and accessible way into a new subject. They are written by experts, and have been translated into more than 45 different languages.

The series began in 1995, and now covers a wide variety of topics in every discipline. The VSI library currently contains over 650 volumes—a Very Short Introduction to everything from Psychology and Philosophy of Science to American History and Relativity—and continues to grow in every subject area.

Very Short Introductions available now:

ABOLITIONISM Richard S. Newman
THE ABRAHAMIC RELIGIONS
 Charles L. Cohen
ACCOUNTING Christopher Nobes
ADAM SMITH Christopher J. Berry
ADOLESCENCE Peter K. Smith
ADVERTISING Winston Fletcher
AERIAL WARFARE Frank Ledwidge
AESTHETICS Bence Nanay
AFRICAN AMERICAN RELIGION
 Eddie S. Glaude Jr
AFRICAN HISTORY John Parker and
 Richard Rathbone
AFRICAN POLITICS Ian Taylor
AFRICAN RELIGIONS
 Jacob K. Olupona
AGEING Nancy A. Pachana
AGNOSTICISM Robin Le Poidevin
AGRICULTURE Paul Brassley and
 Richard Soffe
ALBERT CAMUS Oliver Gloag
ALEXANDER THE GREAT
 Hugh Bowden
ALGEBRA Peter M. Higgins
AMERICAN BUSINESS HISTORY
 Walter A. Friedman
AMERICAN CULTURAL HISTORY
 Eric Avila
AMERICAN FOREIGN RELATIONS
 Andrew Preston
AMERICAN HISTORY Paul S. Boyer
AMERICAN IMMIGRATION
 David A. Gerber
AMERICAN LEGAL HISTORY
 G. Edward White

AMERICAN NAVAL HISTORY
 Craig L. Symonds
AMERICAN POLITICAL HISTORY
 Donald Critchlow
AMERICAN POLITICAL PARTIES
 AND ELECTIONS L. Sandy Maisel
AMERICAN POLITICS
 Richard M. Valelly
THE AMERICAN PRESIDENCY
 Charles O. Jones
THE AMERICAN REVOLUTION
 Robert J. Allison
AMERICAN SLAVERY
 Heather Andrea Williams
THE AMERICAN WEST Stephen Aron
AMERICAN WOMEN'S HISTORY
 Susan Ware
ANAESTHESIA Aidan O'Donnell
ANALYTIC PHILOSOPHY
 Michael Beaney
ANARCHISM Colin Ward
ANCIENT ASSYRIA Karen Radner
ANCIENT EGYPT Ian Shaw
ANCIENT EGYPTIAN ART AND
 ARCHITECTURE Christina Riggs
ANCIENT GREECE Paul Cartledge
THE ANCIENT NEAR EAST
 Amanda H. Podany
ANCIENT PHILOSOPHY Julia Annas
ANCIENT WARFARE
 Harry Sidebottom
ANGELS David Albert Jones
ANGLICANISM Mark Chapman
THE ANGLO-SAXON AGE
 John Blair

Available soon:

For more information visit our website

www.oup.com/vsi/

Jaboury Ghazoul

ECOLOGY

A Very Short Introduction

OXFORD
UNIVERSITY PRESS

OXFORD
UNIVERSITY PRESS

Great Clarendon Street, Oxford, OX2 6DP,
United Kingdom

Oxford University Press is a department of the University of Oxford.
It furthers the University's objective of excellence in research, scholarship,
and education by publishing worldwide. Oxford is a registered trade mark of
Oxford University Press in the UK and in certain other countries

Published in the United States of America by Oxford University Press
198 Madison Avenue, New York, NY 10016, United States of America

British Library Cataloguing in Publication Data
Data available

Library of Congress Control Number: 2020936021

ISBN 978-0-19-883101-3

Printed and bound by
CPI Group (UK) Ltd, Croydon, CR0 4YY

For Dave, the most enthusiastic natural historian I know, and for Sanna, a dolphin scientist in the making

Contents

List of illustrations

Ecology

Chapter 1
What is ecology?

What does that eat?

'What does that eat?' This was, a few years ago now, the constant
refrain of my 3-year-old son, who was so fascinated by the life
around him. It was a question he asked of any animal that
happened to stumble across his field of vision. Frustratingly
repetitive for a parent who wished to imbue more creativity in his
young son's mind, this simple question nonetheless lies at the
heart of ecological science. Ecology deals with how organisms
interact with each other and their environment. This includes
what they eat, and what eats them. Predators, prey, plants,
parasites, pathogens—all use different food sequestering strategies
to obtain energy to survive and reproduce. These different
strategies give rise to patterns in nature (Figure 1). Ecology is, at
its most fundamental, a science that seeks to understand the
biological processes that determine patterns in the natural world.

Ecology is, of course, much more than just patterns structured
by consumption. Organisms compete for scarce resources, and
cooperate for mutual benefits. They change the environment
around them, creating new arenas for interaction, and giving
expression to new patterns in nature. They are constrained by
the surrounding environment, which is complex and changes

1. **Patterns are generated by ecological interactions. In Namibia, 'fairy circles' of bare soil edged by vegetation are thought to be created by a combination of termites that clear vegetation from around their nests, and by plants competing for water.**

across space and in time. Humans too shape ecological patterns and processes, by altering environments and the abundances of organisms.

In my children's early years we used to visit a small pond that, in summer, was festooned with dragonflies and damselflies. The 'dragonfly pond' caught my sons' attention. They recognized many different insects flitting across the surface of the water, an insight into biodiversity that required no textbook introduction. Having determined what these creatures ate, their next question was, inevitably, 'what's that one called?' The celebrated physicist Richard Feynman was adamant that 'names don't constitute knowledge'. That might well be true in the strict sense of the meaning of knowledge, but my children knew better. Curiosity about plant and animal names encouraged them to seek out differences among species, by which they could know that they were, indeed, different species demanding different names. More questions followed: 'What does that one do?', 'Why is this one always in woodland, but that other in meadows?' They were beginning to recognize order and pattern in nature. Ecology begins with such patterns. Patterns make the 'what does it eat?' questions ecologically interesting. Taxonomy, concerned with describing, identifying, and classifying organisms, provides the framework by which ecological patterns and interactions can be recognized and understood. Names open up new vistas of observation and enquiry. Richard Feynman was a superb physicist, but would have been a lousy ecologist.

The Silwood Park campus of the University of London's Imperial College where I worked has extensive lawns much frequented by rabbits in the summer. The kids loved to chase these rabbits, albeit ineffectually. They stopped short, however, when they glimpsed a fox skulking in the undercover at the woodland edge. Foxes were a rare sight that instilled a little anxiety in the children. They knew that foxes ate rabbits but, they wondered, given so many rabbits why do we not see more foxes? Explaining that a fox needs a lot of rabbits to raise a fox family introduced a fundamental law in ecology, that the available biomass, or mass of living organisms, decreases as we move up a food chain, starting with plants, to herbivores that feed on plants, to the predators that eat the herbivores. The ability of consumers to build biomass is limited by their ability to obtain food, and by the efficiency with which they can convert food energy into biomass. That is why there are far fewer foxes than there are rabbits.

Not every summer delivered such profusion of rabbits. In some years, rabbits were decidedly few and far between. The kids began to notice similar year-to-year fluctuations in the abundance of voles, acorns, and beech nuts. Some years delivered a bonanza of apples in Silwood Park's small apple orchard. In other years there were so few fruit that, to secure a decent share, the children had to be sure to get to the trees well in advance of the live-in postgraduate students. The kids were now asking questions about resource fluctuations and population dynamics. Or, why do bees visit the apple flowers? What do worms do? Why do hedgehogs only come out at night? Why do sycamore seeds spin? Why do apple trees make apples? All these questions are ecological.

What is ecology?

Ecology is a science, a topic of disciplinary study. It is also a worldview that emphasizes environmental connectedness, and has become more or less synonymous with 'environmentalism'. Two interpretations of ecology, scientific and cultural, create confusion in how one relates to the other, but also allow ecologically inspired ideas to proliferate through societal discussions. In consequence, ecology is one of the most pervasive sciences in socio-political and cultural narratives. Ecological thinking pervades romanticism, spiritualism, literature, and politics. It has become the motivational driver of modern lifestyle choices and political agendas.

As a scientific discipline, ecology deals with interactions among organisms and their environment. Ecology seeks to describe these patterns, and understand the processes that give rise to them. It is often easy to describe patterns in nature. It is well known, for example, that the number of species increases as we move from polar to tropical latitudes. Understanding the causes of such patterns is more difficult. Some theories seek to link species richness with the abiotic environment, to energy availability, temperature, or precipitation. Other theories emphasize biotic interactions that promote species coexistence. Diseases or predators might disproportionately attack commoner species, or perhaps rare species have specialized life strategies that facilitate their persistence in a crowded and competitive environment. Either way, processes that favour rare species will tend to support larger numbers of species.

Ecology is closely tied to the unifying framework of evolution, and evolution is fundamentally the product of ecological interactions. Stephen Jay Gould's series of essays, published as *Reflections in Natural History*, is a reflection on the interplay of ecology and evolution. Gould himself was not much interested in ecology, perhaps because as a palaeontologist his curiosity was piqued by the historical development of natural systems, and he saw little historical explanation in ecological processes. Despite Gould's lack of interest, ecology does have a historical perspective, as noted by the 19th-century geologist Charles Lyell. Lyell's geology is avowedly historical, founded on observable natural processes of elevation and erosion. Applying this historical perspective to the living world, Lyell argued against a static and largely ahistorical 'balance of nature', and in favour of continual disruption and change shaped by ecological processes of dispersal, predation, and competition. This opened the door to a more dynamic interpretation of the natural world, inspiring Charles Darwin and Alfred Russel Wallace, among others, to develop evolutionary theory through ecological insight.

Ecology only really makes sense in the light of evolutionary theory. Ecological outcomes are essentially evolutionary processes in real time. The persistence of a species is a function of how its individuals interact with those of other species and the environment. Drawing on an oft-used theatrical metaphor, the environment is the stage upon which interactions unfold. Natural selection is the director of an evolutionary play. Ecology is the performance.

Physics envy

Pierre-Simon Laplace, the French mathematician and physicist, argued that it is theoretically possible to know the future of every atom if only we had complete understanding of the current world and all its processes. Physicists are, of course, now fully aware that randomness and probability are unavoidable facts of nature, and chance (stochasticity) is also engrained in ecological theory. Ecological laws are more probabilistic than deterministic.

We might project how populations disperse into available space, given information on species traits, environmental conditions, and resource availability, but we cannot say precisely where, when, and by which individuals the dispersal process will unfold. Ecological laws are founded on probabilistic interpretations of nature, modelled statistically. A strong mathematical tradition in ecology has generated many insights in how populations and communities operate, and yet mathematical models in ecology have far less precision than in physics. This reflects the importance of historical contingency in determining ecological and, arguably, evolutionary outcomes. Processes and patterns in ecology are shaped as much by legacies of what existed before as they are by current ecological processes.

Scientific reductionists argue that by investigating the properties of the component parts of any system, we can understand how the whole functions. While ecologists embrace reductionism in their science, they also recognize that it falls short of providing a complete understanding of ecological systems. What makes biological systems interesting is their 'emergent' complexity. An individual organism is a complex functional unit with properties that are more than the sum of its cells or organs. An ecological system has, similarly, emergent properties derived from interactions among myriads of organisms and species that give rise to complex outcomes, born of processes such as reproduction, predation, competition, mutualism, dispersal, and growth. Moreover, biological processes interact across spatial scales. It is this interaction of parts and processes across scales that gives ecology its most pronounced characteristic, that of a 'holistic' worldview wherein many aspects of a given system need to be considered to understand its emergent properties and outcomes.

Theory of ecology

Ecology has been said to be replete with concepts but devoid of principles. This is slightly unfair, but reflects the difficulty of developing a rigorous and predictive theory based on universal laws in a discipline that is, fundamentally, contingent on past events and perturbations. Despite this inherent indeterminacy in ecology, it is nonetheless possible to outline several fundamental propositions that underpin the science of ecology, and upon which theory can be built.

Most obviously, the heterogeneous distribution of organisms underlies emergent patterns of nature. Species and individuals are not distributed evenly in space and time. Seaweeds and encrusting animals on a rocky seashore, for example, occupy distinct height bands above the low water tide mark. These vertical zonation patterns are outcomes of both biotic interactions among species, and species responses to the physical environment. Biotic interactions can be between individuals of the same species (intraspecific) or between different species (interspecific). They can be antagonistic or mutually beneficial.

Species respond to variations in environmental conditions created by physical processes, be they wave action and salt inundation on coastal shores, declining temperatures along a mountain slope, or increasing seasonality with latitude. Such environmental heterogeneity provides the basic template for biotic heterogeneity. Yet ecological outcomes are sensitive to the contingencies of chance events (a seed lands in one place and not another), and initial starting conditions. Nature, in consequence, is highly dynamic, and ecological prediction is hostage to historical contingency.

Across this dynamic biophysical environment, resources are limited and finite. Resources might be limited by physical processes, such as rainfall regimes limiting water availability, or by organismal exploitation of those resources. The attributes of species, and their strategies to acquire resources, determine the survival and reproduction of organisms in given environments, and consequently their relative abundance and distributions.

7

Finally, evolutionary change is driven by natural selection, which is essentially an ecological process; and evolution shapes the attributes of individuals and species, which determine their ecological properties.

Ecology as a worldview

In our current age of environmental degradation, ecology is the scientific lens by which we can understand the functioning of our natural and agricultural systems on which our own future wellbeing depends. Many professional ecologists are strongly motivated by a desire to improve environmental stewardship, and as Aldo Leopold wrote in a 1947 letter, 'we simply cannot call ourselves ecologists and be indifferent to the slaughter of the biota now becoming worldwide'. Yet ecological science itself is not normative—there is no 'ought' or 'should'. While ecology is not synonymous with environmentalism, it does offer much to environmental management and conservation. These disciplines draw on ecological concepts and theories in pursuit of strategies relating to how we 'ought' to manage ecosystems, resources, and biodiversity. This normative perspective distinguishes these disciplines from strict ecological science. Applied ecology lies between the two, in that it evaluates the consequences of human activities on ecosystems, and explores potential solution options. It is the process of decision-making, rather than the science itself, that makes the normative stance explicit.

A constant irritant to professional ecologists is that a large section of the public equate ecology with environmentalism, and even tree hugging, green spirituality, or, worse still, hippyism. Ecology in the public mind has become much more than the science. It has insinuated itself in the politics and culture of modern society, by which its meaning has been stretched and reshaped. Ecology, adopted and reinterpreted by a variety of subcultures, has influenced and even subverted mainstream culture. Its prevalence in marketing and advertising shows that neither scientists nor environmental activists have a monopoly on ecology as a concept or meme. Cultural interpretations and usage of ecological concepts is a broad and fascinating topic for enquiry, touched upon in the penultimate chapter, but it is the science of ecology to which this book is primarily devoted.

Chapter 2
The dawn of ecology

Throughout history people have been fascinated by the workings of nature, and have drawn on a rich culture of descriptive natural history to relate to the environment around them. Charles Elton, a pre-eminent ecologist of the modern era, described ecology as 'a new name for a very old subject. It simply means scientific natural history.' Some fundamental ecological concepts are discernible in writings on natural history as far back as the classical era. The development of ecological science from these beginnings has mostly been a product of conceptual advances that, over the past century, have created a discipline rich in technical and mathematical complexity. Yet the processes and outcomes with which ecology is concerned remain accessible and understandable to any perceptive observer, facilitated through a frame of natural history that blends insight with fascination.

Classical ecology

Herodotus, who died around 425 BC, is claimed as the father of history, but he was also a keen natural historian. He noted how Nile crocodiles allow birds to forage unmolested on parasitic leeches within their open jaws, a beneficial interaction for both species that we now call a mutualism. A little later, around 380 BC, Plato fulminated against the loss of forests and the subsequent erosion of soils in Attica, giving an environmental aspect to ecological processes. Neither Herodotus nor Plato went so far as to develop an ecological philosophy, and their observations are firmly rooted in natural history.

The first real tremors of ecological thought, at least in Europe, have their origins with Theophrastus, a student of Aristotle. Aristotle himself had made proto-ecological statements in that he acknowledged the relation between animals and their environment, in much the same vein as the earlier observations of Herodotus and Plato. Theophrastus, on the other hand, developed a far more complete ecological interpretation of plants in *De historia plantarum* and *De causis plantarum*. Theophrastus' treatment of the nature of plants had three aspects. The first was the intrinsic nature of a plant, which today we might refer to as its attributes as determined by its genes. The second was the nature of the environment within which the plant occurs, which might or might not favour the plant given its attributes. The third aspect was human agency, which might shape plants independently of their intrinsic attributes or environment. Unlike earlier philosophers, Theophrastus maintained that the goal of living organisms is to produce seed to perpetuate themselves, rather than to provide humans with food, fuel, or other values.

Theophrastus noted that plants only flourish in places suited to their intrinsic attributes. This is comparable to the modern ecological concept of 'niche'. He recognized plants that were adapted to different conditions of aridity, moisture, salinity, and soil type. Different plant species thrive subject to interactions between the environment and plants' inherent tendencies. Some plants, he observed, only thrive in a narrow range of favourable conditions, and have correspondingly narrow distributions. Only some trees are able to grow in mountains, but Theophrastus also recognized that even within mountains there is variation in species type and form, depending on local conditions.

A first approximation of the concept of competition among species is also traced to Theophrastus. He noted that trees growing close together compete for water and light and thus become tall and slender, while trees in more open conditions are not so. Some trees, such as almond, are 'bad neighbours' by suppressing the growth of others. Like Herodotus, he recognized mutualistic interactions, describing how jays bury acorns that later germinate, and how birds spread mistletoe seeds. Like Plato, Theophrastus decried environmental degradation caused by overexploitation of land and forests. Drainage and deforestation, he argued, caused a cooling of local climates and induced soil infertility, and he advocated land management by limiting the harvesting of timber trees.

The ecology of Theophrastus differs from modern ecology in that it lacked any concept of the network of interactions among organisms in a complex community. Neither did he consider population growth and decline, surprising given that Aristotle had earlier described rapid growth and then crash of rodent populations. This is perhaps due to Theophrastus' primary interest being with plants. Still, he had nothing to say about the sequential development of plant communities (natural succession in modern ecological parlance). But to dwell on his omissions would be churlish, particularly as there was hardly any advance on his ideas for another two millennia.

Theophrastus' most important legacy was to attribute to plants a 'purpose' independent of humans. He was consistent in describing nature as a relationship between organisms and their environment. Apart from giving us the first ecological text, we can legitimately attribute to Theophrastus the word 'ecology' itself. His use of the Greek *oikeios*, the descriptive form of *oikos* (house), provides us with the root of 'ecology' as coined in the 19th century by Ernst Haeckel who, being steeped in classical literature, was surely aware of Theophrastus' work.

Systematic ecology

In the centuries since Theophrastus, it is difficult to glean much in the way of an ecological discipline derived from observation. As far as we can tell, even natural history became intertwined with fable. Eventually, in the 17th century, John Ray set aside fable and mythology to apply his keen powers of observation to explain how nature works. Ray's *Catalogus plantarum circa Cantabrigiam nascentium* (A catalogue of plants that grow around Cambridge), published in 1660, noted the habitats (the physical places in which species live) of each of 558 plant species, including bogs, woods, meadows, and riverbanks, and included observations on their biology. He explained how growth rings in ash trees (*Fraxinus excelsior*) relate to the age of the tree, and how the growth of elm (*Ulmus procera*) is influenced by prevailing winds. He determined, correctly, that rape (*Brassica rapa*) and wild turnip (*Brassica napus*) are related on account of a caterpillar species that does not discriminate between these two food plants even while it 'scorns' many others. Ray's later work included studies of birds, fish, and insects. Although his writings are not strictly ecological in the modern sense, in that they are not embedded within an overarching theoretical framework, they did provide solid ground for natural history based on direct observation and deduction.

John Ray's works directly influenced Carl Linnaeus, whose *Systema naturae* (1735) introduced the two-part naming system for all organisms that we use today. This system arranges species into genera and species, and names them accordingly. There are, for example, thirteen species of solitary wasp judged sufficiently similar by taxonomists to be grouped into the genus *Mellinus*. One of these species, named *Mellinus arvensis* by Linnaeus, occurs widely in the UK but is recognized and identified as such in places as distant as Nepal where it also occurs. Linnaeus' universal classification system names species unambiguously, and this allows for the rigorous study of species, their interactions, and their distributions.

A notable milestone along the road of ecology's emergence is marked by the 'parson-naturalist' Gilbert White, and his *Natural History and Antiquities of Selborne* (1789). The book is a compilation of letters, ostensibly to other naturalists though never posted, comprising observations on the natural history of plants and animals in the vicarage of Selborne in southern England. Gilbert White's observations were meticulous and detailed and, importantly, undertaken in nature itself. White was able to distinguish three near-identical birds, the chiffchaff, willow warbler, and wood warbler, as three species based on their different songs. His *Natural History* includes hundreds of observations on the dates of the seasonal appearance of migratory birds, which provide a valuable baseline to compare against increasingly early appearances of migratory birds in the current warming climate. White recognized an interdependence among organisms that sustained the natural world he observed around Selborne. He described ecological processes such as pollination and seed dispersal, and noted the importance of earthworms as a 'small and despicable link in the chain of nature, yet, if lost, would make a lamentable chasm'.

Humboldt's physical picture

Neither Ray's compendium, Linnaeus' classification system, nor White's collected 'letters', are true works of ecology in the modern sense. Their observations interpreted the natural world with innovative clarity, but remain a far cry from the pursuit of a causal interpretation of processes and patterns, guided by theoretical principles. It was the advent of global scientific exploration in the late 18th century that proved formative. Many state-sponsored excursions to far-flung regions of the world (often thinly veiled expeditions of colonial conquest), as well as impressively adventurous individual explorers, advanced Western science by accruing specimens, observations, and ideas. Back in Europe the foundations of ecology as a science distinct from natural history began to take shape as returned adventurers mingled with, or became, armchair theorists.

Alexander von Humboldt was among the first to evaluate the relationship between organisms and their environment in his botanical geography. His *Essai sur la géographie des plantes* (*Essay on the Geography of Plants*) published in 1807 described the distribution of animals and plants in relation to physical conditions of temperature, altitude, humidity, and atmospheric pressure. At the back of the *Essay* a large fold-out *Physical Picture of the Andes and Neighbouring Countries* depicts species' distributions across a section of South America from the Pacific coast lowlands, across the Andean range (and specifically the Chimborazo volcano), and on to the edge of the Amazon basin (Figure 2). With this *Physical Picture*, Humboldt established the idea that different plant species occupy distinct climatic zones. Patterns in plant distributions could now be investigated and understood in terms of biogeophysical conditions.

Charles Darwin was inspired by Humboldt's science and sense of adventure, but the geologist Charles Lyell introduced the young Darwin to the conflict that lies at the heart of Darwin's evolutionary thinking. Charles Lyell provided the intellectual bridge from Humboldt to Darwin by presenting a view of nature that stood in marked contrast to Gilbert White's homely Selborne, or Humboldt's static *Physical Picture*. In the second volume of *Principles of Geology* (1832), Lyell emphasized the prevalence of predation and competition, a 'struggle for existence', among organisms—even 'the most insignificant and diminutive species, whether in the animal or vegetable kingdom, have each slaughtered their thousands'. Lyell was himself inspired by Augustin de Candolle, who in 1820 had written, 'Toutes les plantes d'un pays, toutes celle d'un lieu donné, sont dans un état de guerre' ('All the plants of a country, in a given place, are in a state

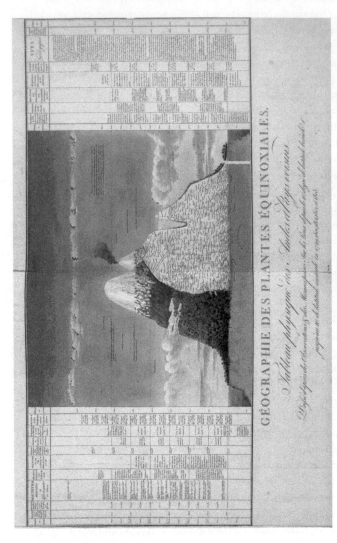

2. Humboldt's *Physical Picture of the Andes and Neighbouring Countries* (here shown in the German version) published in *Essay on the Geography of Plants*, 1807, was the first substantial description of patterns in species distributions based on biophysical attributes of geography.

of war'). Alfred Tennyson described this view of nature as 'red in tooth and claw' (*In Memoriam*, 1850). Lyell's ecological insights, largely overlooked by modern ecologists, anticipated several issues of current interest in ecological research, including ecological cascades ramifying through biological communities.

Darwin was also much influenced by Malthus' (1978) writings on human population growth. Malthus assumed exponential growth of populations would rapidly exhaust resources, which would invoke the competition among individuals that lies at the heart of Darwin's theory of natural selection.

Ecological communities

Plants and animals were beginning to be recognized as belonging to distinct communities. In 1825, the naturalist Adolphe Dureau de la Malle (1777–1857) referred to communities of co-occurring plant species as a *societé*. On broad geographic scales, Augustin de Candolle, in his monumental and unfinished *Prodromus Systematis Naturalis Regni Vegetabilis* (*Natural History of the Vegetable Kingdom*) recognized that plants have particular geographies, which he attributed to temperature. Wladimir Köppen used this idea for his climate classification, which explicitly linked seasonal precipitation and temperature patterns to vegetation formations such as tropical rain forests.

In 1877 Karl Möbius described the detailed interactions between different organisms in an oyster bed in the Bay of Kiel, to which he applied the term 'Biocoenosis', meaning the assemblage of plants and animals and their interactions at a particular location and time. Ernst Haeckel, understanding Darwin's fundamental message, recognized that we must study the entangled network of connections between organisms and their physical and biotic environments if we are to assess attributes that underpin their success. In 1866, Haeckel postulated two sub-disciplines of evolutionary science, ecology (*Oecologie*) and biogeography, which

he called *Chorologie*. In his 1869 inaugural lecture at the
University of Jena, Haeckel presented an eloquent definition of
ecology embedded within evolutionary thought: 'By ecology we
mean the body of knowledge concerning the economy of
nature—the investigation of the total relations of the animal
both to its inorganic and to its organic environment…in a word,
ecology is the study of all those complex interrelations referred
to by Darwin as the conditions of the struggle for existence.'
A further two decades would pass before *ecology* would catch on
as a term. Its first appearance in a book title was not until 1885, in
Hanns Reiter's *Die Consolidation der Physiognomik als versuch
einer Oekologie der Gewaechse (A Physiognomic Synthesis as an
Attempt for an Ecology of Plants)*.

Ecology remained largely descriptive until Eugen Warming began
to consider how abiotic factors such as drought, flooding, fire, salt,
and cold, as well as herbivory, affected the assembly of biotic
communities. By studying plant morphology, Warming began to
explain how species were adapted to the environmental conditions
in which they occurred, and why unrelated species that occupied
habitats with similar abiotic conditions often shared similar traits.
He drew his observations widely, from his native Denmark to
northern Norway and Greenland, and from the Brazilian cerrado.
Warming's *Plantesamfund—Grundtræk af den økologiske
Plantegeografi* published in 1895 (translated into English in 1909
as 'Oecology of Plants') had a profound impact on later British
and North American ecologists, including Arthur Tansley, Henry
Cowles, and Frederic Clements. The American Henry Cowles was
so taken with lectures based on Warming's book that he taught
himself Danish to read the original text prior to the availability of
a translation. Cowles later became known for his own work
(published 1911) on the sequential development of ecological
communities (ecological succession) in sand dune systems of
northern Indiana. Cowles acknowledged the earlier and
related work of Adolphe Dureau de la Malle, and of the Finnish
botanist Ragnar Hult who published the first comprehensive

study of ecological succession in 1881, recognizing that early colonizing plant species form a 'pioneer' plant community, which is gradually replaced by a smaller number of species in more stable communities.

In the first half of the 20th century, ecological thinking was dominated by Clements's climax theory of plant community development. Drawing on observations of the prairie vegetation of Nebraska and the western United States, Clements presented the idea of 'succession', in which plant communities develop in a directional and predictable sequence of stages towards an ultimately stable 'climax state' that is best suited to the local conditions. In *Plant Succession* (1916) Clements argued that certain groups of species were always associated together. The species depended on the group, and the group on its component species, in much the same way that an animal and its organs are co-dependent. The Clementsian approach of a community as a distinct unit was challenged by Henry Gleason, who viewed vegetation as a continuum, not a unit, with associations being merely coincidental. His 'individualistic concept of ecology' placed much greater weight on the attributes of individual species being the main determinants of community structure, with plant associations being much less deterministic and structured than Clements's theory allowed.

While individualistic perspectives have come to dominate modern ecology, descriptive classifications of vegetation associations remain useful. Mapping vegetation types across Britain, championed by Arthur Tansley among others, became an important purpose of ecological study in the first half of the 20th century. This eventually led to the current British National Vegetation Classification (NVC), a comprehensive description and classification of 286 plant communities across 12 major vegetation types, ranging from forests to grasslands, wetlands, coastal communities, and heaths. The NVC provides a widely accepted standard for understanding community associations across Britain, and one that is recognized by forestry, agricultural, and conservation agencies, as well as governmental and corporate organizations. National vegetation classification systems have now been developed by many other countries to provide a basis for ecological and biodiversity studies, conservation assessments, and the planning of ecosystem management and restoration work. Despite the inherent complexity and contingency that underpins ecological communities, such pragmatic classifications provide a common language for interpreting the different communities in a region.

Systems thinking

Following Clements and Gleason, Arthur Tansley in 1935 contributed, arguably, the next major conceptual advance in the development of ecological science. Tansley, enthused by Warming's book, led the Central Committee for the Survey and Study of British Vegetation to coordinate ecological studies across the country. Tansley's interactions with European and American ecologists, including Carl Schröter in Switzerland, and Henry Cowles and Frederic Clements in America, began to establish ecology as an international discipline. Tansley's great contribution was the ecosystem concept. In a well-known 1935 paper on vegetation concepts, he argued that organisms and environment should not be treated separately but be considered together as 'one physical system', an 'ecosystem'. Ecosystems integrated the biotic community with the physical environment to form 'recognisable self-contained entit[ies]' which are, according to Tansley, the basic units of nature. Clements's concept of an interdependent community of organisms was, by Tansley's reckoning, incomplete without inclusion of transfers of energy and materials between organisms, and between organisms and their environment. Tansley was immensely influential far beyond the British ecological spectrum, but he is particularly highly regarded in the UK also through his role in establishing the

British Ecological Society in 1913 as the first professional society of ecologists. Tansley was its first president.

The ecosystem concept initially gained little traction. Raymond Lindeman developed the concept by focusing attention on flows of energy across trophic levels or ecosystems compartments. He recognized that only a small fraction of organic energy is transferred from one trophic level to the next, with around 90 per cent of energy consumed being lost through respiration or incomplete digestion. This '10 per cent law' explains Charles Elton's observation of declining number or biomass of organisms at higher trophic levels. Lindeman would, no doubt, have elaborated these ideas further were it not for his early death at only 27 years of age.

The ecosystem concept began to gain wider traction when Paul Richards argued in 1952 that it is 'preferable to regard soil, vegetation, animal life, climate and parent rock as components of a single system, the ecosystem'. Other ecologists such as Eugene Odum and Robert Whittaker also acknowledged ecosystems as fundamental organizational units that encompass interdependent relationships, food chains, physical processes, and regulatory pathways. They began to interpret ecosystems through attributes such as energy flows, productivity, dynamics, and disturbances that transcended individuals or species. Mathematical modelling began to be used to simulate ecosystems, while experimental manipulation of ecosystems was increasingly adopted to understand the causal relations underlying ecosystem processes and outcomes. These approaches helped to steer ecology from a largely descriptive to a more predictive science.

Animal ecology and Elton's niche

While much of early ecology evolved from the study of the distributions and associations of plant species and communities, the ecology of animals was following a somewhat different and parallel path. Animal ecology owes its early development to Charles Elton. Elton immersed himself in natural history, being inspired by his older brother Geoffrey. There is likely to have been little natural history to study in industrial Manchester, his home city, at the turn of the century, but Geoffrey and Charles were lucky to enjoy family holidays in the rural Malvern Hills in Worcestershire. Elton turned his childhood interest into a fruitful career in ecology, starting with a survey of the fauna of Spitsbergen in 1921. On the return journey from a 1923 expedition to the Arctic, Elton read *Norges Pattedyr*, by the Norwegian biologist Robert Collett, which described population explosions, migrations, and mass drowning of lemmings. Elton recognized the dramatic population fluctuations as characteristic of animals in the Arctic. This idea ran counter to the prevailing assumption that populations persist in a state of balance. Moreover, understanding why populations fluctuate could uncover the mechanisms that regulate populations. In 1925, the Hudson Bay Company hired Elton to work on the population fluctuations of snowshoe hare and Canadian lynx, these being relevant to the company's fur harvests. His studies of lynx–hare population oscillations have since become a staple of undergraduate ecology courses.

Elton's *Animal Ecology* (1927) founded the discipline. It explained how a small number of principles governed the structure and function of animal populations and communities. This included their arrangement in food chains that took the form of a 'pyramid of numbers', where a large biomass of plants supports a smaller biomass of herbivores, which in turn supports smaller masses of predators (Figure 3).

Animal Ecology also introduced the 'niche' concept, describing how animals are adapted to and constrained by their community, particularly in relation to food availability and predators. A niche in common parlance is akin to an alcove, a small space tucked away in some building or in the corner of a room. Settling into a

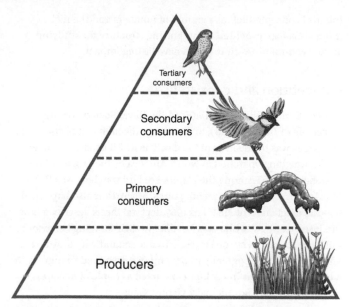

3. The trophic pyramid, representing the transfer of energy from plants, as primary producers, to herbivores, and predators. Roughly 10 per cent of the energy available at one level is transferred to the next, resulting in a declining abundance or biomass at higher trophic levels.

niche conveys a sense of security and comfort. Elton adopted this term for ecology, by applying it to the particular conditions and resources to which a species is adapted, and where it can thrive and reproduce. Charles Elton described the niche as an organism's mode of life 'in the sense that we speak of trades or jobs or professions in a human community'.

Elton was adamant that ecology is a field science, 'scientific natural history' as he called it, where careful observation of animals in their habitats will uncover the rules of nature. Observations on feeding by individual animals can reveal insights on population size and animal community structure within the frame of concepts such as pyramid of numbers and the niche. *Animal Ecology* provided an organizing structure for studying animal communities that was to have lasting impact.

Competition and coexistence

In 1889 Emily Williamson founded the Royal Society for the Protection of Birds (RSPB), to discourage the killing of birds for hat plumes, as was the fashion of the day. The RSPB ushered in an era of bird watching, matched in North America by the growth of the Audubon Society. Among the plethora of bird watchers was Robert MacArthur. Apart from watching birds, MacArthur also happened to be an excellent ecologist. His 1957 doctoral thesis described how five insect-eating warbler species differ in their foraging behaviour in North American spruce forests. To understand why they do so, and why this is ecologically important, we first need to digress to a classic experiment in ecology conducted a couple of decades earlier by the Russian biologist Georgy Gause.

In 1932 Georgy Gause published *Experimental Studies on the Struggle for Existence*. This grandly titled paper described a simple but elegant series of experiments in which Gause monitored populations of two closely related species of *Paramecium*, single-celled organisms that feed on bacteria and yeast. Both species thrived when grown under identical conditions in separate containers. But when the two species were grown together, the population of one grew rapidly at the expense of the other, which was eventually eliminated from the container. Gause had shown that two species requiring the same limited resource cannot coexist: the better competitor excludes the other. This was later coined the *competitive exclusion principle*. The principle had earlier precedents. In 1904 Joseph Grinnell described how two species must differ in some traits relating to their productivity in order to coexist, but it took Gause's experimental confirmation to embed the concept into mainstream ecology.

To coexist, competition must be avoided. Gause repeated his experiment with a different *Paramecium* pair combination. On this occasion, both species persisted. Closer inspection revealed that one *Paramecium* species tended to feed on bacteria suspended in the culture medium, while the other fed on yeast at the bottom of the culture tube. By specializing on different resources, or by developing different strategies for obtaining resources, the two *Paramecium* species were able to avoid competition and thus coexist in the same habitat.

Robert MacArthur's five warbler species appeared to contradict the competitive exclusion principle. In the breeding season, these very similar birds occur together in spruce forests where they feed on the same insect prey. Through diligent observation, MacArthur discovered that each bird species feeds in a different position on the tree, and adopts different foraging behaviour, resulting in different prey selection (Figure 4). By segregating the place and manner of feeding, these five bird species differentiate feeding niches, which enables them to minimize competition, thereby permitting coexistence.

| Cape May warbler | Black-throated green warbler | Yellow-rumped warbler | Bay-breasted warbler | Blackburnian warbler |

4. **Resource partitioning of five warbler species in the white spruce forests of Maine in North America. The diagrams illustrate the positions on spruce trees where birds did most of their foraging.**

Watching shorebirds on a coastal beach reveals another example of niche differentiation, with bird species using different feeding strategies to target different prey items on different parts of the shore. At the top of the beach plovers run up and down sandy areas feeding on small arthropods, while turnstones flip shells over and rummage through seaweed searching for crustaceans. Low down on the shore, long-billed curlews probe for burrow-dwelling crabs and shrimps, while heavy-billed oystercatchers pry open clams and mussels at the tide's edge. On the mid-shore between these two groups, sandpipers pick up small worms and arthropods stranded by the receding tide. With competition thus avoided, coexistence is possible.

Evelyn Hutchinson, in his *Concluding Remarks* (1957), probably the most opaque title for a paper in the history of ecological science, formalized the ecological niche of a species as a multi-dimensional 'parameter space', with each dimension representing a resource. Water availability, temperature, or light, represent just a few of many different niche dimensions that collectively define where species can live, grow, and reproduce. Predators and competitors preclude species from occupying the full range of their environmental niche space. Species might also fail to occupy the entirety of their potential geographical niche space because they have not been able to disperse into and colonize all liveable regions. On this basis, Hutchinson distinguished between the *fundamental* niche, which describes the complete possibilities of a particular species' occurrence, and its *realized* niche, the more limited range of conditions in which it actually persists, given competitors and predators, and the contingencies of dispersal.

Ecologists demand experimental confirmation that observed behaviours conform to explanatory theory. We owe one of the best niche studies to a 'Scottish landlady, Mrs Plant, whose generous terms for board and room' enabled Joseph Connell, a young American ecologist in the 1950s, to extend his stay on the Isle of Cumbrae in the Firth of Clyde (apparently her wonderful soup helped ward off the Scottish weather). After two unproductive years trying to catch rabbits in the Californian Berkeley hills, Connell vowed never to work on anything larger than his thumb, and so turned his attention to barnacles. On Cumbrae, he noted that of the two barnacle species commonly found along the coast, the large *Semibalanus* occurs low down on the shore, while smaller *Chthamalus* is confined to the higher intertidal regions that are regularly exposed by the receding tides. By experimentally excluding *Semibalanus*, Connell revealed that *Chthamalus* larvae grow and thrive in lower shore regions. Normally, though, *Semibalanus* excludes the smaller barnacle by smothering or undercutting it. *Chthamalus* survives in the upper shore because *Semibalanus* is unable to tolerate extended periods of desiccation when the receding tide exposes the shore. Joseph Connell's work is seminal as he showed, using field experiments, how species niches are constrained by combinations of biotic and abiotic factors.

A balance of nature?

An apparent stability or orderliness about the organization of the biological world is an idea that stretches back to Herodotus, and earlier in eastern philosophy. On the surface, it seems that nature is, indeed, more or less in equilibrium. Populations fluctuate but nonetheless persist over long periods. Ecosystems appear to last even when subject to periodic disturbances, and show an inherent tendency, a resiliency, to recover following perturbation. While a balance of nature is imbued with mythological and cultural significance, and ecosystems seem stable enough to the casual observer, ecology as a modern science demands empirical evidence that is guided by, and contributes to the development of, an ecological theoretical framework.

Elton's 1958 book *The Ecology of Invasions by Animals and Plants* argued that simple communities are less stable than complex ones. To support this thinking he noted that pest outbreaks occur most frequently in simple agricultural systems rather than more complex natural systems, or in simple temperate forests rather than species-rich and complex tropical forests. Although ecological stability and its relation to ecosystem complexity has been much debated since, Elton's lasting contribution to ecology, among many, has been his community-oriented thinking, involving many species interacting over large temporal and spatial scales.

Robert MacArthur, writing at the same time as Elton in the 1950s, explained stability in terms of the fluctuations of species' population abundances in relation to each other. Stability, he argued, is when species within the community maintain steady populations despite fluctuations in the populations of other species. This most likely occurs when species interact with many other species, as when a predator feeds on many prey species, such that a substantial decline in one species has little effect on others. A large number of weak interactions appear to stabilize fluctuations in hypothetical food webs, but real food webs have many other attributes that seem to confound the emergence of clear complexity–stability relationships.

By the mid-20th century ecological thinking began to coalesce around three broad areas, each linked to the apparent stability and persistence of species, communities, and ecosystems. The first attempted to understand how natural populations are regulated. The second sought to understand how communities are organized, particularly in terms of successional processes, food webs, and interactions among species. The third explored how energy moves across trophic levels or ecosystem boundaries. These broad themes provided a loose disciplinary framework, but ecology quickly flourished into multiple fields of interest, some focusing on fundamental understanding of ecological processes and the nature of stability, and others emphasizing how these processes underpin conservation, sustainable futures, and human wellbeing.

In the 1970s, Robert May applied non-linear mathematical models and deterministic chaos to ecology, to tackle the essence of ecological complexity and stability. He showed that, contrary to common intuition, complex ecosystems are less likely to be stable than simple ones. While the complexity–stability debate remains unresolved, non-linear concepts and models are now central to ecological thinking, and link enquiries on the nature of stability to those concerning our sustainable future. Ecologists warn of thresholds and tipping points when natural systems experience sudden and dramatic changes in form and function, often driven by environmental changes or biodiversity losses caused by human actions. Understanding the underlying drivers and causes of tipping points, at least sufficiently to predict them, is one of ecology's current foremost challenges.

Chapter 3
Populations

Explosions of lemmings

Lemmings are, reputedly, collectively suicidal. Populations of these small rodents of the Arctic tundra explode every few years, only to collapse dramatically, often within months (Figure 5). When conditions are favourable, lemmings achieve reproductive maturity at less than two months, and most females reproduce several times through the summer season, on each occasion producing litters of up to six young. It is no wonder that such phenomenal reproductive capacity drives explosive population growth. Predators, including weasels, snowy owls, and arctic foxes, multiply quickly during such periods, but not sufficiently quickly to control lemming numbers. What ultimately drives lemming populations into steep decline is not predation, but exhaustion of their food. Lack of food often drives lemmings to undergo mass migrations in search of better pastures, and high mortality during such migrations has earned them their suicidal reputation. Rather than suicide, lemmings are victims of poor foresight.

It is not only lemmings that have boom and bust population cycles. Any keen gardener must be vigilant to the sudden appearance of aphids and voles in numbers that quickly undermine months of hard work. Sudden eruptions of swarms of desert locusts in Africa have biblical fame (Box 1), and can have serious economic and

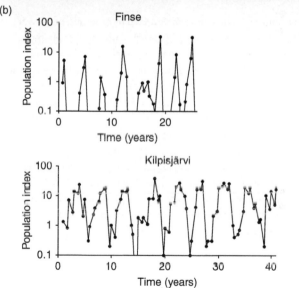

5. (a) A lemming. (b) Population dynamics of lemmings at Finse, Norway, and voles at Kilpisjärvi, Finland, showing outbreak events and subsequent population collapses.

Box 1 Locust plagues

Desert locusts occur from Mauritania to India, usually in low numbers. They increase rapidly, however, following good rains and subsequent growth of fresh vegetation. Large aggregations of young wingless hoppers and winged adults form within two or three months, often across areas of about 5,000 km². If rains continue, locusts move into adjacent areas of fresh vegetation and multiply through several successive generations. The immense upsurge in numbers creates swarms that engulf whole regions. In favourable conditions a plague develops. Plagues can number up to 150 million locusts per square kilometre, and in one day a 1 km² locust plague can eat the same amount of food as 35,000 people.

While outbreaks are common, few lead to upsurges, and fewer still create plagues. The condition that allows such phenomenal population growth is the continued availability of fresh food over large areas, which is dependent on extensive rains. Heavy rains in East Africa in late 2019 gave rise to a huge locust swarm in January 2020 that destroyed crops in Somalia, Ethiopia, and Kenya. As I write these words, at the end of January 2020, there is concern that seasonal rains in March will stimulate new vegetation growth across much of the region. This could cause the numbers of fast-breeding locusts to multiply 500-fold, before drier weather in June limits their spread. Until 2020, the last major desert locust plague was in 1987–9, while a later upsurge in 2003–5 affected the whole Sahel, from Senegal and Mauritania to the Red Sea.

human impact. Similarly, rodent outbreaks in Asia cause an annual loss of rice that might otherwise feed about 200 million people.

Weather and food are common causes of population outbreaks. Abundant rainfall drives vegetation growth and seed production, to which lemming, locust, or mice populations respond. Mild winters and warm springs contribute by reducing winter mortality, giving populations a boost at the start of the breeding season. This has been the cause of mice outbreaks in places as disparate as California, Hawaii, and Australia. Mild winters are also behind immense bark beetle outbreaks that are currently destroying huge areas of North American forests.

Outbreak species are characterized by very high *intrinsic growth rates* (r), the rate of natural increase that a population has the potential to achieve. Characteristic traits of such species are early and frequent reproduction, with many offspring in each reproductive episode. Actual population growth rates are often far lower than the intrinsic growth rate, as the majority of young die before becoming reproductively mature. But if environmental conditions are favourable, and on occasion they are, many of the offspring will survive and reproduce. Rapid population growth ensues, driving outbreaks. Outbreaks quickly exceed the available resources, and in short order the population crashes. That is why we are not inundated with lemmings.

Predator control

The vole, another small Arctic rodent, has population cycles characterized by gentle rises, prolonged peak abundance phases, and declines that are more gradual than those of lemmings (Figure 5(b)). During peak population phases, female voles mature more slowly, and their reproductive rates decline. Voles eat grasses that recover quickly, and so food is less limited. Whereas lemmings catastrophically exhaust their food supply, the lower reproductive rate of voles maintains peak population abundance over several years. A rise in predator populations eventually drives the decline of the vole population.

Aldo Leopold argued eloquently in *Thinking Like a Mountain* (A Sand County Almanac, 1949) that the extirpation of wolves from a landscape is akin to have given 'God a new pruning shears, and forbidden Him all other exercise'. His argument was that having removed wolves, deer will multiply to an extent that all bushes and seedlings would be browsed 'first to anaemic desuetude, and then to death'. Wildlife managers use this reasoning to argue that wolves, lynx, mountain lions, and other large predators are needed to control grazing animals for the wider benefit of the ecosystem (Box 2).

Box 2 Landscapes of fear

For those of us accustomed to a safe and tranquil environment, stumbling across a wild wolf or bear releases a primal feeling of fear. We take our subsequent steps in the wild rather more carefully, and proceed with heightened alertness to our surroundings. Prey species, constantly in uneasy company with their predators, most likely feel the same way. Herbivores are more wary and skittish in the company of predators. Predator tracks, smells, growls, and occasional glimpses create a constant state of vigilance and apprehension. A 'landscape of fear' forms in the minds of prey, which distinguish habitats and locations by their relative risk or safety.

This landscape of fear, created by the mere possibility of predator presence, can be ecologically more meaningful than any direct predation. Following the reintroduction of wolves to Yellowstone National Park in 1995, after a sixty-year absence, the resident elk population declined steeply. The decline far exceeded what could be attributed to predation alone. Elk spent more time on the move and, with less time to forage, could only raise a third as many offspring. Yellowstone has since experienced an increase of willow and aspen trees, released from the pressures of elk herbivory, and beavers are resurgent following the recovery of their food trees.

Debate continues as to the extent to which changes in the Yellowstone ecosystem are attributable to the return of the wolf. An experimental demonstration of the landscape of fear, enacted on the rocky shores of the Gulf Islands of British Columbia, is less ambiguous. Using a speaker system, raccoons were exposed to either the sound of barking dogs (which kill raccoons) or seals (which don't). On hearing dogs, raccoons became more vigilant and spent less time foraging along shorelines. Rock pools on these shorelines experienced dramatic increases in their fish, worm, and crab biota.

It is, however, difficult to determine whether predators are controlling prey populations, or if prey populations, controlled by some other factor, regulate the predators. A classic data set, derived from records of pelts received by the Hudson Bay Company, reveals cyclical oscillations of snowshoe hare and its predator the Canadian lynx (Figure 6). For many years, predation by lynx was thought to reduce the abundance of hares, which in turn collapses the lynx population. The subsequent release from predation pressure allows hares once more to grow in number, to which the lynx population responds, initiating the cycle once again. Yet snowshoe hare populations follow similar ten-year boom-and-bust cycles on islands that have no lynx. It now seems more likely that the periodic crashes in hare populations occur when large numbers of hares exhaust their food plants. Following the crash, plants slowly recover and the hares increase again. The lynx population might merely track the hare population.

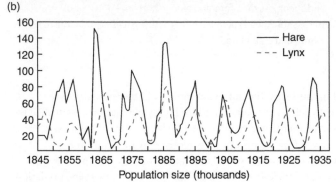

6. (a) Canadian lynx and snowshoe hare. (b) Population cycles of the snowshoe hare and Canadian lynx over a 100-year period, based on numbers of pelts supplied to the Hudson Bay Company.

Competitive regulation

Unlike lemmings, voles, and hares, most species maintain relatively stable populations, despite inherent capacities to increase exponentially. In 1954 David Lack, a British ecologist and evolutionary biologist, described it thus, 'most wild animals fluctuate irregularly in numbers between limits that are extremely restricted compared with what their rates of increase would allow'. Few species have the high intrinsic growth rates of lemmings and voles, and any population change will therefore be more gradual. Predators also play their part in controlling population numbers, but a more pervasive control mechanism is competition.

Competition occurs when there is not enough of a necessary resource to meet the needs of all individuals in the population. As a population grows, so does the density of its individuals. At low densities resources are plentiful, and individual reproduction and survival are high. Population growth can approach the theoretical maximal rate of increase per individual, the intrinsic population growth rate (r). As density rises, average per capita resource availability declines, and individuals begin to compete for limited resources. Organisms that lose out have fewer offspring or die young, and this slows population growth. Eventually, available resources might become so limiting that mortality across the population exceeds births, and the population begins to decline. Resource availability per individual, manifested through density-dependent competition, regulates population size.

Competing individuals need not interact directly. Exploitation of a limited resource by one individual reduces its availability to other individuals even though these organisms might never meet. Similarly, a plant might deplete soil nutrients to the detriment of neighbouring plants. Nevertheless, organisms often do interact directly in a competitive arena. Some animals actively exclude others from access to resources by defending an exclusive territory. Many birds, mammals, fish, and even insects defend territories to secure exclusive access to nesting sites, feeding areas, or mates. Defending a territory can be costly. It incurs constant vigilance and energy, and risks injury or death. Animals therefore only hold and defend territories when resources are scarce, as there is little point in defending resources that are widely abundant.

r and K strategies

Density-dependent processes tend to regulate populations around an equilibrium point where birth rates equal death rates. This equilibrium point is the environment's *carrying capacity* (*K*), the population size that can be maintained given the available resources. Populations tend to increase towards *K* as they utilize available resources, but decline on exceeding *K* as resources become too limited to support all individuals in the population. Populations therefore oscillate around the carrying capacity (Figure 7). Species that have high intrinsic growth rates often have large amplitude oscillations, as they tend to greatly overshoot *K* and suffer large population crashes in the aftermath. Populations with low *r*-values grow more slowly, and track *K* more closely. The carrying capacity itself might change according to environmental conditions.

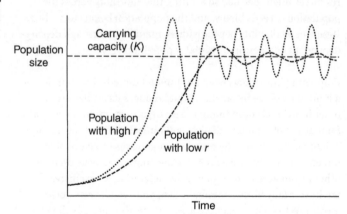

7. **Idealized population growth and oscillation around the carrying capacity. Species with high intrinsic growth rates (*r*) are characterized by population overshoot followed by die-offs, and these are referred to as *r*-selected species. Species with low *r* track the carrying capacity (*K*) more closely and experience much less population fluctuation. These are *K*-selected species.**

Ecologists envisage two generalized life-history strategies based on the intrinsic rate of population growth (r), and the carrying capacity (K). Species that are r-selected reproduce early, produce many offspring, and undergo rapid population growth. Lemmings are r-strategists. So are many garden weeds. They perform well when resources are plentiful. Resources are usually locked up in biomass, or preferentially captured by highly competitive species, so r-selected species often depend on disturbances to make resources available once again. Disturbance can take the form of a falling tree that creates a gap in the forest canopy allowing light to flood the dark understorey, or a fire that returns nutrients locked-up in vegetation back to the soil. The sudden abundance of light, nutrients, and other resources allows r-selected species, with their rapid reproductive rates, to quickly dominate new space by exploiting the abundant resources. As resources decline, r-selected species are gradually replaced by species that are more competitive in capturing the now scarce resources.

K-selected species occupy habitats where there is less environmental fluctuation, and fewer disturbances. Large mammals, many birds, and forest trees are K-strategists. These species tend to have longer lifespans, allowing them to defer reproduction to later in life when they are larger and more competitively secure. Offspring are comparatively few, but tend to be better provisioned (with large seeds, for example), or are invested with extended parental care, and offspring survival is therefore high relative to r-selected species. Populations of K-selected species generally remain close to the environment's carrying capacity and, in the absence of large disturbances, they maintain fairly stable populations.

The intrinsic rate of population growth and the carrying capacity have direct practical importance: they are used to quantify extinction risk in conservation contexts, model commercial fisheries, or assess potential growth of pest invasions. In practice, it is very difficult to specify growth rates and carrying capacities for species in particular environments, or to predict patterns of population recovery. This is because resources fluctuate, populations are affected by competition with other species, and by predation, and forces such as climate and disturbances affect populations irregularly and at different spatial and temporal scales.

Deterministic chaos

When intrinsic growth rate values are very high, we enter the realm of chaotic dynamics, characterized by large and erratic population fluctuations. Long-term prediction of chaotic population dynamics is not possible, despite the simplicity and determinism of the underlying population growth model, largely derived from the intrinsic growth rate. Chaotic systems are very sensitive to initial conditions, and slight differences in the population size, or indeed in the estimate of the intrinsic growth rate, become magnified to create large differences in outcomes.

Chaotic population dynamics driven by deterministic processes initially attracted much theoretical interest in the 1970s and 1980s, as a few simple population dynamics equations appeared sufficient to explain seemingly complex dynamics. Predicting dynamics that are chaotic remained impossible, yet just knowing that population fluctuations are deterministic (rather than truly random) indicates the existence of an underlying mechanistic process, the understanding of which can provide insights into ecological processes. Laboratory studies provided empirical confirmation of chaotic dynamics in beetle and plankton populations, among others, and it seemed that chaos theory might explain many ecological patterns.

Enthusiasm quickly waned, however, when it proved difficult to relate deterministic processes to ecological dynamics outside tightly controlled laboratory conditions. For one thing, random 'environmental noise', such as weather variation, droughts, or hurricanes, repeatedly impacts population growth, but these chance events were largely excluded by both theory and laboratory studies. Moreover, the attraction of theoretical deterministic processes was their 'low dimensionality', or the limited number of parameters required to reproduce population dynamics. Yet most ecological communities comprise populations of many species, most interacting weakly and some interacting strongly. Competitive and predatory interactions between species tend to dampen population fluctuations and therefore limit the emergence of chaotic behaviour. Consequently, the most likely candidates for chaos in natural systems are those with tightly linked species–resource interactions little affected by other competing species or predators. Lemmings, for example.

Rock pools in New Zealand provide an interesting case of chaotic multi-species population dynamics in a natural system. In these rock pools, barnacles colonize bare rock surfaces. As they grow they provide conditions that are suitable for settlement by crustose algae and mussels. The mussels eventually smother and kill the underlying barnacles. Dead barnacles, and their overlying mussel mats, are easily detached from the rock, thereby exposing bare rock surface once more. The cycle from bare rock, to barnacles, algae, mussels, and back again to bare rock, is a deterministic process that sustains all three species, but none is able to stabilize its population. Modelling shows that in the absence of seasonal weather, such a system is likely to settle into a stable and predictable coexistence of the three species. The system is, in fact, rather less predictable, as warm summer days cause high mortality among mussels and crustose algae, which introduces chaotic fluctuations in species abundances. In this example, a deterministic sequence is made chaotic by seasonal weather that differentially affects survival of two of the three species.

Life history trade-offs

The r and K dichotomy is a simplification of the wide range of strategies that plant and animal species exhibit. These strategies are reflected by species traits that include gestation length, age of reproductive maturity, litter size, frequency of reproduction, and maximum lifespan. Collectively, these attributes describe the main demographic aspects of a species, or its 'life history'. As we have seen, many species respond opportunistically to favourable periods by growing rapidly, and reproducing early and prolifically. Others grow slowly, and produce only a few offspring at a time. We might expect profligate species to quickly outnumber and displace the more conservative species. This does not happen for two main reasons. One we have already come across—profligate species quickly exhaust their resources so that density-dependent processes limit reproduction and survival.

Another reason is the cost of reproduction or, more precisely, the trade-offs between growth, survival, and reproduction. A 'trade-off' is a zero-sum game, or a direct negative relationship between two attributes in which an increase in one is associated with a decrease in another. Given finite resources, the more resources invested in the production of offspring the less are available for growth. The small width of growth rings of temperate trees, such as oak or beech, during years of high seed production bears witness to this. Similarly, fish that defer reproduction to later in life, such as sharks, attain large body sizes relatively quickly, as resources are allocated predominantly to growth. Good gardeners will know that pruning the ripening seed heads of perennial plants will improve survivorship, growth, and flowering in the following year.

There are many life-history trade-offs. Another is offspring production set against their survival. A large investment to each offspring, be it in terms of gestation, food, or parental care, increases the chances of its survival to maturity, but limits the total number of offspring that can be supported. Orchids produce as many as three million tiny dust-like seeds in each fruit capsule, but allocate no food resources, and survival depends on securing a connection to specialized fungi that feed the seedlings. The vast majority of these seeds fail in this quest and quickly die. By contrast, the coco-de-mer palm has the largest seed in the world, weighing up to around 15 kg. These well-provisioned seeds germinate and grow for months without drawing on anything other than the seeds' own resources, but the mother tree produces only one seed per year.

Alternative strategies allocate different resource ratios to growth, maintenance, and reproduction to suit environmental circumstances. These trade-offs are also expressed among individuals within a species. In many species, the trade-off between reproduction and growth determines body size, which has implications for survival. Guppies, small fish familiar to many aquarium owners, develop different allocation strategies depending on the predators they are exposed to. In Trinidad, the native range of guppies, some streams harbour cichlid fish that feed on large guppies, while in other streams killifish feed on small guppies. In cichlid-dominated streams, guppies allocate more resources to reproduction early in life, and therefore reproduce at a smaller size. In streams dominated by killifish, guppies delay reproduction and prioritize resource allocation to growth, achieving larger body sizes more quickly, thereby lessening the risk of death by killifish.

Functional traits

Functional traits are aspects of organisms' physiology (metabolic rate, frost tolerance, or photosynthetic rate), morphology (beak size, body mass, leaf area, or wood density), or behaviour (feeding or predator evasion strategies), that influence performance or fitness. Functional traits are aligned to life-history strategies, either as evolved characters, or as morphological or physiological trade-offs. A young tree might allocate most of its resources to growth, a life-history strategy, at the cost of fewer resources being available for defences, a functional trait, to protect against herbivory. Such a strategy is successful where resources are plentiful, as any tissue lost to herbivores can be readily and rapidly replaced. In less favourable habitats, perhaps a forest understorey where photosynthetic light is scarce, a plant with this strategy is unlikely to survive long as replacing lost tissue is very slow in the low light conditions.

Functional traits influence the abilities of species to colonize or thrive in a habitat, and to persist in the face of environmental changes. They also affect ecosystem properties. Plants in habitats of low soil fertility or little rainfall tend to have small thick leaves, with high mass relative to surface area, to reduce water loss and improve nutrient use efficiency. Such leaves decompose more slowly, and nutrient cycling rates are slower in these communities. In aquatic habitats, predation pressure tends to favour larger planktonic organisms that have some degree of protection by virtue of size, but large size increases sinking rate of dead plankton, and thus the rate of nutrient transfer to the sediment, which influences biogeochemical cycling. On large scales, the ecosystem effects of functional traits have important consequences for our own societies that depend on nutrient and biogeochemical cycles to replenish soil fertility and marine fisheries, or to sequester and store atmospheric carbon.

Dispersal

We have considered population dynamics as the outcome of births and deaths regulated by intrinsic density-dependent processes. Population dynamics are also affected by the spatial distribution of populations in the landscape, and the dispersal of individuals across the landscape, and from one population to another. If a population occupies a safe and favourable habitat, there might seem to be little motivation for its individuals to disperse out of it. This then begs the question as to why dispersal has evolved at all, and why does almost every species have dispersal strategies.

Chance alone suggests that over the long term, any isolated population will be driven extinct by environmental change or disturbance. Dispersal allows individuals to colonize other patches, thereby lessening the probability of extinction of the population as a whole. Dispersal also enables recolonization of a patch from which a population was previously extirpated.

Independent of any environmental change, a growing population occupying a finite patch becomes increasingly subject to density-dependent competition as resources are consumed. Dispersal is a means of escaping density-dependent constraints. Emigration not only increases the survival probability of the emigrant (assuming density-dependent costs outweigh those of dispersal), but also reduces the density of the source population, thereby increasing individual reproduction probabilities. The reproductive rate of voles, for example, increases if the local population size is alleviated by dispersal of some of the population. Selection therefore favours dispersal as a strategy for individual (and hence population) survival both in the short and long term.

Through dispersal, individuals might also escape the pests and diseases that become prevalent in large populations. In Paris, for example, butterflies of the Pierid family colonize isolated habitat patches in the city centre, but the parasitoid wasp that feeds on the butterfly larvae cannot reach these city centre localities from the Parisian periphery (Figure 8). Similarly, the survival of tropical

8. **Tiny parasitoid wasps are unable to disperse into the centre of Paris, and so Pierid butterflies are free from attack in the city centre.**

tree seedlings often depends on the successful dispersal of seeds away from the parent tree where high numbers of pathogens and herbivores reside.

Metapopulations

Landscapes can be considered to contain discrete habitat patches of variable quality. Some patches can only support small populations, which are vulnerable to local extinction. Immigration from good-quality habitats can prolong the viability of these marginal populations. High-quality 'source' patches support large populations from which there is net emigration, and smaller or lower-quality 'sink' patches can only support small vulnerable populations that are maintained by immigration, or become extinct and are subsequently re-established through immigration. This source–sink 'metapopulation' model emphasizes the importance of dispersal in maintaining populations across a landscape.

Environmental change, often caused by humans, can degrade habitat patch quality, which undermines metapopulation structure by turning source populations into sinks. The complete loss of habitat patches, perhaps by their conversion to other uses, can increase the distance between remaining patches. This can make it more difficult for dispersing organisms to find favourable patches, consequently reducing the likelihood of population rescue by immigration. For this reason, conservationists argue the need to create habitat corridors or 'stepping stones' to facilitate the dispersal of animals (and seeds) across a landscape from one favourable habitat patch to another.

Dispersal also facilitates the exchange of genetic diversity across a metapopulation, helping to overcome problems associated with inbreeding. Inbreeding, which arises from repeated mating between genetically similar individuals, can increase the prevalence of genetic ailments and reduce survival probabilities. Isolated populations are more likely to suffer from inbreeding, especially if descended from a small number of colonizing individuals without subsequent immigration. On the Åland islands in Finland, the Glanville fritillary butterfly occurs in a metapopulation of hundreds of variably sized but discrete populations. The smallest and most isolated populations tend to have low genetic diversity, and are prone to extinction due to inbreeding. Less isolated patches avoid inbreeding by regular exchange of individuals.

Managing populations

Managing species, be it for our own resource requirements, for conservation, or some other purpose, relies on our understanding of population dynamics, as shaped by species traits and life-history strategies, environmental conditions, biotic interactions, and dispersal. Managing the expansion of problematic invasive species or pests, or conserving small populations of endangered species, requires a good understanding of species traits, and how they relate to population growth rates, density-dependent competition, and metapopulation dynamics. The coupled dynamics of lynx and snowshoe hares, or the periodic rise and fall of lemming populations, provide insights into how simple predator–prey–resource interactions might regulate animal populations. Yet most interactions are far more complex, involving multiple species across a range of different environmental conditions. Ecology has successfully used simple models to clarify and elucidate basic and general principles, and we use the outputs of these models to manage populations according to our needs. The generic insights gained come at the cost of realism in particular contexts. We should interpret them with caution. We have a long way to go before we understand the dynamics of more complex multi-species ecological systems, at least sufficiently well enough to predict how they might respond to anthropogenic change.

Chapter 4
Communities

Nature, red in tooth and claw?

Alfred, Lord Tennyson's poem, *In Memoriam*, a requiem for his close friend Arthur Henry Hallam, contrasts humanity's faith in love with Nature's savagery:

> Man...
> Who trusted God was love indeed
> And love Creation's final law—
> Tho' Nature, red in tooth and claw
> With ravine, shriek'd against his creed.

Incessant predation and competition appear to justify Tennyson's revulsion of Nature. Yet cooperation among Nature's species is just as prevalent (and of course 'Man' himself is no angel).

Species that cooperate for mutual benefit are called mutualists, and collaborative interactions are mutualisms. A 'symbiosis' (from the Greek for 'living together') is a mutualism where the involved species form close interdependent physical partnerships. Lichens, for example, are symbiotic associations between a fungus and photosynthesizing algae or cyanobacteria. Most mutualisms are far less intimate. Bees fleetingly visit flowers to collect pollen for their brood, and by doing so plants are pollinated and seed produced. Floral nectar has no function other than to entice pollinators to flowers. Similarly, fruits are no more than offerings to animals that eat them, thus dispersing the seeds they contain.

The study of mutualisms has a long history, but less developed theory. 'Mutualism' was first used as an ecological term in Pierre-Joseph van Beneden's 1875 book *Les Commensaux et les parasites (Animal Parasites and Messmates)*. A few years later in 1878, Heinrich de Bary, known to students and colleagues alike simply as 'The Professor', introduced 'symbiosis' as a biological concept in his lecture *Die Erscheinung der Symbiose (The Phenomenon of Symbioses)* to the Association of German Naturalists and Physicians. The Professor defined symbiosis as 'a phenomenon in which dissimilar organisms live together', which encompassed parasitism. By the century's end, many mutualisms were recognized. Yet through the 20th century, there was little effort to develop a theory of mutualism. Many early scientists who wrote about mutualisms apparently had left-wing sympathies, and some have suggested that an association with left-wing politics might have curtailed interest in mutualism theory. If true, then the publication in 1902 of *Mutual Aid: A Factor in Evolution* by the anarchist Peter Kropotkin most likely played a part. Even van Beneden's first reference to 'mutualism' in 1875 could be an allusion to the 'Mutualité' societies of workers in France and Belgium established in the early 19th century to provide mutual financial aid.

Cooperation

Coral reefs, extending over thousands of square kilometres of tropical oceans, are among the most biologically diverse ecosystems on Earth (Figure 9). They would not exist were it not for a partnership between the coral (an animal) and certain species of photosynthetic dinoflagellates, a group of otherwise mostly free-living planktonic organisms. The symbiotic dinoflagellates, known as zooxanthellae, are housed by the coral

9. A healthy coral reef ecosystem depends on an interdependent relationship between an animal, the coral, and a dinoflagellate protist.

and gain nutrients from the prey that corals catch, for which they pay rent to the coral with photosynthetically derived carbohydrates. Zooxanthellae provide up to 95 per cent of the coral's carbon, which enhances coral calcification and allows construction of the massive carbonate reefs. Zooxanthellae species have different photosynthetic capacities under different light conditions, and corals replace some zooxanthellae with others as environmental conditions change. Recent high sea surface warming events, likely linked to global warming, can result in wholesale rejection of all zooxanthellae in many corals. This causes coral bleaching, and even coral death if the zooxanthellae are not replaced quickly. By disrupting the coral–zooxanthellae symbiosis, climate change threatens the integrity of coral reefs, and the rich biological communities they support.

The coral–zooxanthellae symbiosis underpins the growth of coral reefs, but their persistence depends on a more diffuse mutualism between corals and algae-eating fish. Reef fish remove around 90 per cent of algal production, which maintains a clean reef surface suitable for settling coral larvae and coral health. Reef fish, in turn, benefit from the nooks and crannies of the coral reef, which provides food and protection from predators.

Another symbiotic relationship that forms the foundation of an entire community occurs around hydrothermal vents in volcanically active areas of oceanic crust. Emanating from these fissures are highly acidic superheated fluids, as hot as 400°C, rich in hydrogen sulphide, toxic to most life. Yet the biomass of animals in the immediate vicinity of these seemingly hostile environments is some 1,000 times that of the surrounding ocean plain. Much of this biomass is of giant bivalves and tubeworms that, curiously, have neither mouth nor gut. They acquire nutrition from sulphur-oxidizing bacteria living within their bodies. The animals provide carbon dioxide, oxygen, and hydrogen sulphide to the bacteria, which synthesize organic compounds that are absorbed by the hosts. This symbiosis establishes a food web, supported entirely by bacterial chemosynthesis in these lightless depths, which includes a variety of crustaceans, anemones, and several fish and octopus species.

Many other less intimate partnerships have similar importance for ecosystem structure. Whistling-thorn acacia (*Vachellia drepanolobium*) constitutes 95 per cent of woody cover in East African black-cotton savannah. Elephants feed on tree bark, leaves, and twigs, and inflict catastrophic damage to trees in general, but the whistling-thorn acacia remains unmolested. The acacia has bodyguards, in the shape of four ant species. These ants nest in the acacia's hollow swollen thorns and feed on nectar secreted at the base of its leaves (Figure 10). The ants aggressively bite the sensitive interior of the trunk of any elephant foolish enough to feed on acacia branches. Without these bodyguards, acacia trees of black-cotton savannahs would also be destroyed by elephants, and the wooded savannah replaced by fire-maintained open grasslands, which support far fewer elephants.

10. Whistling-thorn acacia trees host ant bodyguards that defend them from herbivores, and provide the ants with a home and with sugary nectar in return.

Uneasy relationships

For all Kropotkin and van Beneden's admiration of mutualism in the natural world, nature is, in reality, rather more fickle. A mutualism is only a mutualism if neither partner is able to exploit the other, and mutualistic partners are rarely equal. One species often benefits more from the interaction than the other, and relationships change according to circumstance. A cooperative partnership can quickly descend into an exploitative one.

Consider the ant bodyguards of the whistling-thorn acacia. From the ants' perspective, energy invested by the tree on flower and seed production is energy lost to the creation of hollow thorns in which the ants nest, and sugary nectar. The ants will consequently nip many of the flowers off the tree as it attempts to flower. Conversely, if herbivores are scarce, the acacia no longer needs such strong protection, and produces fewer thorns and less nectar, to the detriment of the resident ants. The ants retaliate by raising sap-sucking insects, from which they harvest honeydew and so obtain their carbohydrate needs. Mutualisms often oscillate between mutually beneficial partnerships and opportunistic parasitism.

Many plants have 'mycorrhiza' fungi that form a sheath around the roots, or penetrate roots and root cells. The fungi increase the efficiency with which the plant is able to access scarce soil nutrients, notably phosphorus and nitrogen. The fungal mycelium, composed of tiny filamentous structures called hyphae, can extend to a length of 100 m within one cubic centimetre of soil, accessing soil pores that plant roots are unable to reach. In return, the fungi receive carbohydrates that plants generate through photosynthesis. Plants might allocate as much as 20 per cent of photosynthetically derived carbon to fungal partners. This mutualism is predicated on the basis that the benefits gained by each partner outweigh the costs they each incur. The cost to the plant is the carbohydrate allocated to the fungus that might instead have been used for plant growth. Usually, this cost is worth bearing to access scarce nutrients, but in fertile soils plants might have little, if any, need of mycorrhizal fungi. The fungi continue, nonetheless, to draw on the plant's carbohydrates, which reduces plant growth. The relationship has turned exploitative, or even parasitic. Plant–mycorrhizal interactions range from strongly mutualistic and interdependent, to weakly mutualistic, weakly parasitic, or even strongly parasitic, depending on the 'balance of trade' of costs and benefits mediated by soil fertility or light availability for photosynthesis. Such shifts in relative costs and benefits between mutualistic species are commonplace.

Partnerships are challenging, but three's a crowd. Plants support many mutualistic partners, including mycorrhizal fungi, pollinating insects, and seed dispersing vertebrates. The plant side of the bargain is to provide carbon, in roots for fungi, as nectar for pollinators, or fruit for seed dispersers. If soil fertility is high, the plant has less need for the mycorrhizal fungus, and can instead allocate more carbohydrates to flowers and fruit to enhance seed production. Yet perfect redistribution of resources is often not possible due to physiological constraints, and fungi continue to draw at least some carbohydrates from the plant, potentially limiting its reproductive potential. Differentiating mutualisms from parasitic relationships can be difficult when resource trade-offs among multiple partners are involved.

Fidelity and infidelity

Two species depending completely on each other are vulnerable to the decline of either one. Most mutualisms are, therefore, rather diffuse, as each species engages with several partners. This spreads the risk of the loss of one or more partner species across the network of partners. Plants, for example, have a broad suite of mycorrhizal fungal partners, and each mycorrhiza fungus associates with many plant species. Similarly, most plants receive many different pollinating insects, just as these same pollinating species visit many different flowering plants.

Highly specialized mutualisms do, nonetheless, exist. As anomalies, they are the focus of considerable ecological interest in that they reveal the underlying conflict that characterizes mutualisms. Each fig tree species, and there are roughly 750 of them, is pollinated by one, or very few, species of tiny fig wasp. Female wasps, barely one millimetre in length, lay eggs in the tiny flowers enclosed within the fig. Wasp larvae parasitize the fig by feeding on plant tissues that might otherwise have formed seeds. Having completed their development, a new generation of adult fig wasps mate and leave the fig, but only after being doused with pollen. They seek out new figs in which to lay their eggs. In doing so they pollinate many of the flowers, and those flowers that escape parasitism develop into seed. There are many variations on this theme. Some fig wasps are true parasites, in that they contribute nothing to pollination.

Fig trees have evolved different floral forms to retain some control over the relationship. Some figs include both short and long flowers. Fig wasps are able to lay their eggs in the short flowers, but are unable to do so in elongated flowers. Other fig species separate male and female flowers on different trees. Figs of male trees contain both male and female flowers, but the female flowers function only to breed pollinators and do not produce seed. On female trees the flowers have a structure that enables the wasps to pollinate but not to lay eggs. As wasps cannot distinguish between fig sexes, pollen-laden wasps entering female figs pollinate the flowers, but fail to reproduce themselves.

Similar conflicts exist among North American yuccas and their pollinating yucca moths that feed on a proportion of yucca seeds, or *Chiastochaeta* flies that pollinate but also feed on ovules of European globeflowers. There is little to separate mutualism from parasitism in these relationships. Both partners depend on each other completely, but underlying this ecological cooperation is evolutionary conflict. Each partner seeks to minimize its costs and maximize its benefits. This plays out in all mutualisms, but it is in the specialist mutualisms where the conflicts inherent in cooperation are laid bare.

Cascades

Some species, while not direct mutualists, nevertheless indirectly benefit other species through their activities. Entire communities might depend on such 'keystone' species, the loss of which has impacts that cascade across the biological system.

In seas too cold for coral reefs, seaweeds provide the physical ecosystem architecture. *Macrocystis* kelps of temperate rocky shores off the west coast of North and South America grow to 60 m, forming great underwater kelp forests that support rich fish and mammal communities (Figure 11(a)). The sea otter maintains this system by feeding on herbivorous sea urchins, allowing the

11. A rich kelp forest community maintained by sea otters (a), without which herbivorous sea urchin numbers get out of control (b).

kelps to flourish. Fur trappers hunted otters to near extinction in the 19th and early 20th centuries, and by the 1920s only small remnant populations remained in Siberia, Alaska, and California. The resulting proliferation of urchins grazed and destroyed kelp forests, creating 'urchin barrens' devoid of other animals or algae

(Figure 11(b)). Urchins, which can live for several years on little sustenance, prevented the establishment and recovery of new kelp. It took the reintroduction of sea otters in the 1970s to reduce urchin numbers sufficiently to allow kelp forest recovery. Since 2013 a new crisis has been unfolding off the coast of California. A viral outbreak has almost wiped out many starfish species. Starfish are normally voracious urchin predators, and their loss led to an explosion in urchin numbers in 2014 and 2015, and another collapse of the kelp forest ecosystem. This also means a loss of commercial fisheries, including the red abalone, a delicious mollusc that was, until recently, abundant in the kelp forest systems.

Communities also experience cascading changes when habitats are fragmented, creating small and isolated habitat patches. Small patches support smaller populations, which have higher extinction probabilities. The upshot is that small patches support fewer species. This is why conservationists seek to maintain large habitat blocks. John Terborgh took advantage of rain forest flooding during the infilling of Lago Guri hydroelectric reservoir in Venezuela to document the cascading impacts of habitat fragmentation. As waters rose, forested hilltops became islands isolated from neighbouring hilltops and mainland forest. Terborgh studied changes in animal populations on twelve islands, some small, others large. Within a decade, animal populations had deviated markedly from mainland forests. Capuchin monkeys disappeared on small islands, but persisted on the larger islands. On small islands, bird densities were twice that of mainland populations, but on larger islands densities had declined to one-fifth of mainland populations. Monkeys raided birds' nests on the larger islands, but birds were safe from capuchin predation on small islands. Other changes were more remarkable. Rodent and iguana populations were, respectively, 35 and 10 times higher than mainland populations. Howler monkeys attained densities of 1,000 per square kilometre, far higher than mainland densities of up to 40 per square kilometre. Most remarkable was the explosion of leaf-cutter ants, which were some 100 times as abundant on islands as in mainland areas. The disappearance of top predators, such as jaguars, explained the enormous increase in these herbivorous animals. The resulting increase in herbivory had cascading effects on the plant community by driving higher tree mortality and preventing regeneration by seed. Surviving plants were those that were protected from herbivory by being inedible or toxic, which is likely to have further implications for animal populations as well as processes such as nutrient cycling on these islands.

Succession

An enduring ecological concept, at least since the 1920s, has been that of succession, the sequential development of increasingly complex ecological communities, eventually settling at a stable end-point called the climax. Primary succession occurs on freshly exposed landforms, such as newly formed volcanic flows or emergent islands, or land exposed by retreating glacial ice. Retreating glaciers in the Alps expose bare rock surfaces that are first colonized by short-lived grasses and herbs, followed by perennial herbs and woody bushes, and later still by trees. These stages of succession can be followed by walking from a glacier's snout and down the valley that it previously occupied (Figure 12).

In secondary succession, the vegetation might be partially or even completely removed by fire, pathogen attack, human action, or some other perturbation, but the soil, and probably some seed and remnant vegetation, remains. Secondary succession from abandoned fields has created the mature forests of eastern America (Figure 13). The hills of New England were once busy farming communities that had cleared most of the pre-colonial conifer and hardwood forests. In the 19th century, New England farmers followed trails to new lands on the western American frontier. Their New England farms, no longer economically

12. Succession in Morteratsch valley, Switzerland, where the retreating glacier exposes new land that is colonized by plants and develops, over time, into a more complex and rich community. All three photos were taken from the same position, and where the glacier front was located in 1970, looking up the valley towards the retreating glacier. The top image was taken in 1985, the middle image in 2002, and the bottom image in 2018.

13. A mature but secondary forest (within the Harvard Forest) now covers what were once extensive field systems in rural Massachusetts that were abandoned in the mid-19th century. Old field walls criss-crossing the forest are legacies of past land use.

tenable and now abandoned, were quickly re-colonized by new tenants, beginning with annual weeds and herbaceous perennials, later followed by longer lived shrubs and fast growing trees, quaking aspen and paper birch, and ultimately by the large trees, sugar maples, beeches, hemlocks, and red oaks, that now occupy these former farms. The forests of the eastern United States are almost entirely secondary regrowth and not primary, undisturbed forest.

Does succession end? Left undisturbed, a community might tend towards a state dominated by large trees that cast a deep shade under which only similarly slow growing shade tolerant species are able to persist. This is the climax state, a theoretical construct with a long and contentious history. In 1916, Frederic Clements argued that given sufficient time, a single climax dominates any given climatic region regardless of its environmental starting point. This view, challenged by Arthur Tansley and Henry Gleason, led to a drawn-out and sometimes bitter debate on organizational processes underlying succession. Clements treated communities as super-organisms that undergo a series of developmental stages, each with its own internal organization. Gleason argued that interactions among individual species determine successional sequences, and that outcomes are largely non-deterministic, being influenced by chance events of dispersal and the ability of individuals to colonize, establish, and compete successfully for resources. In contrast to Clements's interpretation of a community as a single 'organic entity', Gleason states that a community 'is not an organism, scarcely even a vegetation unit, but merely a *coincidence*' (emphasis in the original). Tansley argued that a local climax is determined by several factors. Climate is but one; others include soil, geology, aspect, and topography. In practice, it is difficult to identify discrete and stable climax communities, as the structure and composition of communities vary continuously along many environmental gradients. Disturbance is, in any case, always present in one form or another.

The rate of change in a community is, however, often imperceptible, and this is sufficient to declare a climax community has been reached. Succession on an abandoned field might take 100 or 200 years to reach a climax state, as has most likely happened in much of the eastern United States. During this time windfalls, disease outbreaks, and fires reset successional processes. The retreat of northern ice sheets some 10,000 years ago initiated successional processes that are, arguably, continuing to this day, so it is questionable whether a theoretical state exists in reality.

Competition and colonization trade-offs

Why does succession happen, and why does it happen the way it does? Many theoretical explanations have been proposed for the sequential turnover of species, from fast-growing herbs and forbs to slower growing, longer-lived woody shrubs and trees, paralleled by the development of increasingly diverse and complex communities. Many of these explanations revolve around ubiquitous trade-offs in ecology. No single species can excel at everything. As a species adapts to a particular condition, its ability to cope with other conditions diminishes. Ecology appears to be a zero-sum game, and trade-offs loom large in successional processes, including that of competition versus colonization.

Individuals that do well early in successional sequences belong to species that grow rapidly, so long as resources are plentiful. These are short-lived species that colonize bare ground, or fast-growing herbs and shrubs that benefit from resource abundance. Resources are, indeed, abundant early on, as demands on space, light, and nutrients are low, given the initially small community biomass. This changes as more seeds arrive and plants establish. Increasing numbers of plants, all jostling for space, drives competition for diminishing resources. Early arrivals with high resource demands are unable to sustain rapid growth when competing with later arrivals that use meagre resources more efficiently. The slower growing but more efficient plants, in time, shade out and replace the faster-growing resource-demanding species.

The survival of populations of fast-growing species requires that they continue to colonize disturbed sites where there is a relative paucity of competitors and an abundance of resources. Environmental disturbances create such sites, but locating them by seed dispersal is largely a matter of luck. To maximize colonization success, early successional plants use resources profligately to produce great abundances of small seeds. These propagules are widely dispersed, often on wind currents, with a few lucky ones landing in favourable spots of resource abundance. The tiny seeds hold little in the way of reserves for the plant embryos they contain, so they are not likely to survive competition with larger seeded plants unless they happen to land in resource-rich environments. Competitive species, by contrast, tend to produce large seed packed with resources. This gives the young plants a head start in a competitive environment.

Combining a strong competitive streak with reproductive profligacy would seem to be an unbeatable strategy, but this is precluded by trade-offs. Producing large numbers of seed is costly in resources, which diminishes resources allocated to growth, to roots to access nutrients, leaves to capture light, strong woody stems to resist mechanical damage, or to chemical toxins to protect against pests and herbivores. Trade-offs preclude super-plants.

What of animals?

When we think of succession, we mainly think of plants, for the obvious reason that terrestrial community development is largely dependent on the establishment, growth, and turnover of herbs, shrubs, and trees that give structure to the environment and create the habitat in which animals live. Nonetheless, animals can and do influence successional trajectories and outcomes. Vertebrates have crucial roles in changing the relative abundances of seeds through seed predation, and in colonization processes through the seeds they disperse. Vertebrates disperse over 60 per cent of woody plant species in the eastern deciduous forests of the United States, while birds and mammals disperse between 60 per cent and 95 per cent of woody species in tropical and subtropical forests across the world. Larger-bodied animals disperse large-seeded plants, which tend to be later-successional species. Such animals tend to be more vulnerable to hunting and environmental degradation, and their loss would most likely affect the successional pathways of regenerating habitats. The large seeds of many late successional plants are also attractive food packages for hungry rodents. Granivores shape successional processes by preferentially reducing the seed densities of large-seeded plants.

Vertebrates can stop succession altogether. High numbers of red deer in the Scottish Highlands prevent woodland regeneration on extensive areas of treeless heather moorland. Similarly, high numbers of elephants in African savannahs browse, break, and crush trees, favouring the persistence of grasses, which are then maintained by fire.

Ecosystems

The debate among early 20th-century ecologists concerning the succession of plant communities and the nature of the climax is reflected to this day in the tension between the holistic and reductionist approaches to ecosystem ecology. Tansley defined ecosystems as 'including not only the organism complex, but also the whole complex of physical factors forming what we call the environment of the biome'. He emphasized the coupling of biological, chemical, and physical processes into a single ecological system. Some recent definitions capture the spirit of Tansley's interpretation, as 'a unit comprising a community (or communities) of organisms and their physical and chemical environment, at any scale desirably specified, in which there are continuous fluxes of matter and energy in an interactive open system'. 'Desirably specified' ecosystem units can be as small as the community of organisms living in a water-filled cavity of a pitcher plant. More often, ecosystems are delineated at larger spatial scales within relatively distinct environments such as a stream, lake, or woodland. Ecosystem definitions have been muddied by more fluid interpretations that encompass wide swathes of environmental and social science. These latter interpretations are more holistic in their inclusion of people and their actions as integral elements of ecosystems.

Raymond Lindeman interpreted ecosystems through the relationships of biotic and abiotic compartments by which energy and matter flow. The efficiencies of energy capture and retention by these compartments depend on the physical structure and trophic organization of ecosystems. The physical structure of ecosystems includes the size and distribution of physical features of the ecosystem. These might be predominantly non-biological in aquatic systems, deserts, or tundra. In these systems, rocks, sediments, water, or ice primarily limit the distribution, abundance, and complexity of the biota, which has little capacity to modify the environment. In ecosystems that are more productive, such as forests, ecosystem structure is primarily biological. Trees capture solar energy and sequester nutrients thereby substantially modifying the abiotic conditions by, for example, contributing to the formation of soils, contributing detritus to river systems, slowing erosion, regulating temperature and precipitation, and altering disturbance regimes.

Trophic structure

Trophic organization of the ecosystem is characterized by the food web. Ecosystems can be studied in terms of the energy captured from the sun and stored as carbohydrates in plant tissues, which is transferred along various feeding pathways from herbivores to carnivores. Decomposition releases nutrients back to abiotic ecosystem components, while energy is dissipated as heat, and from the combined respiration of the community.

With very few exceptions, sunlight powers Earth's biota. Its capture by plants during photosynthesis forms the basis of the food chain, the first trophic level in any community. The exceptions include chemosynthetic bacteria in deep dark oceans, which produce biomass from the oxidation of hydrogen sulphide or ammonia rather than by photosynthesis. Raymond Lindeman conceptualized food chains and food webs as the transfer of energy from plants, the first trophic level, to subsequent trophic levels of herbivores, carnivores, and decomposers. The available energy at any trophic level is a function of its biomass, the mass of organisms in that trophic level. By examining the transfer of energy across an ecosystem's trophic levels it is possible to determine how much biomass an ecosystem can support.

Fungi and animals, as well as most bacteria, are not able to synthesize new biomass through photosynthesis, and derive their matter and energy needs from plants. They do so by eating plants, or indirectly by eating each other. Primary production is the rate of biomass produced per unit area by plants, the first trophic level. Secondary production is the rate of new biomass production by consumer organisms. Plant eaters constitute the second trophic level, while carnivores feeding on the herbivores occupy the third. There might even be predators that feed on animals in the third trophic level. Very rarely are there more than four trophic levels in any community. The reason is, in part, a matter of energy transfer.

Secondary production by herbivores is around an order of magnitude less than primary production. Similarly, only around one-tenth of herbivore production is converted into predator biomass. The biomass available to organisms occupying higher trophic levels is therefore a tiny fraction of that produced by plants. There is simply not enough energy in higher trophic levels to support viable populations of yet more trophic levels.

What happens to plant biomass and the energy it represents? Most is simply not consumed, and when plants die their biomass (now properly called necromass) is utilized by the community of decomposers in the soil, mostly bacteria, fungi, and various invertebrates, that return the plant material to the soil. Much of the eaten biomass cannot be easily converted to animal tissue, and is instead excreted. Animals are not well equipped to digest the structurally complex carbohydrates (lignin and cellulose) that account for a high proportion of plant tissue. Consequently, they only assimilate around 20 per cent to 50 per cent of biomass they consume, although seed or fruit eaters can assimilate as much as

70 per cent of the energy. Carnivores feeding on animal tissue, by contrast, assimilate around 80 per cent of consumed biomass. Animals also use some of the energy from consumed biomass to do work, be it capturing prey, escaping from predators, pursuing mates, defending territories, building nests, or migrating. Moreover, inefficiencies in the use of energy and its conversion from biomass result in losses in the form of heat. Only a tiny amount of the solar energy captured by plants is therefore used to create animal biomass (Table 1).

It is for reasons of inefficiencies in energy conversion that the biomass and number of higher trophic levels is but a tiny fraction of that of plants at the lowest level. Charles Elton first noted the roughly tenfold decline in biomass from each trophic level to the next, but it was Raymond Lindeman who was able to explain this pattern by energy transfer and losses. There is no escape from the second law of thermodynamics: energy is dissipated as it does work in the conversion of material from one form to another up through the food chain. It is the second law of thermodynamics

Table 1. The biomass of plants, insects, and vertebrates, compiled from a range of ecosystems, shows that only a tiny fraction of plant biomass is converted into animal biomass (all values in grams per square metre)

Ecosystem	Plants	Insects	Vertebrates
Tropical lowland forest in Peru	39,000	5.4	0.15
Temperate coniferous forests	30,000	2.4	0.08
Temperate deciduous forests	20,000	5.0	0.11
Tropical grassland, Serengeti	3000	0.76	2.3
Temperate grassland, Colorado, USA	2300	0.62	1.1
Cropland, Poland	1260	5.8	0.2
Stream, Arizona, USA	350	3.0	50

that explains why, in Paul Colinvaux's words, big fierce animals are rare.

Biogeochemical cycles

Lindeman's emphasis on flows of energy and materials across ecosystems links ecological and biogeochemical processes. Biogeochemical cycling arises from the fluxes of materials among biotic and abiotic ecosystem components. Plants, for example, take up nutrients from the soil, facilitated by the action of soil biota, and by symbioses between plants and mycorrhizal fungi or micro-organisms. Nutrients move up through the food web when herbivores consume plant tissue, and are eventually recycled back into the soil through decomposition. Consumers, be they herbivores or predators, accelerate nutrient recycling by defecating, and redirect nutrient flows by moving across landscapes or between ecosystems. The periodic emergence of masses of reproductive ants or cicadas, or more precisely their subsequent death and decomposition, contributes a nutrient pulse into aquatic ecosystems that is sufficient to stimulate aquatic productivity. Reverse flows occur when adult midges emerge from their larval aquatic habitat, increasing up to fivefold nitrogen and phosphorus inputs to terrestrial ecosystems within 50 m of streams.

Changes in the species composition of communities following disturbance or during succession affect rates and pathways of nutrient flows. Early successional plant communities, which establish soon after a perturbation, when resources are abundant, are generally wasteful of nutrients as there is little or no competition. More nutrients are lost from the system during these stages than in later successional stages when competition imposes greater retention and recycling of nutrients. Nutrients lost from one ecosystem become inputs to another. Detritus washed into streams during storms is the primary source of nutrients for many stream ecosystems, and organic matter carried downstream is a major source of nutrients for estuarine and coastal ecosystems.

Intensive agricultural and silvicultural systems are relatively inefficient at retaining nutrients because the few species they contain cannot capture the different forms of matter as well as more diverse plant communities. Diverse species mixes encompass a wider range of strategies for nutrient capture, and have a greater variety of mutualistic interactions with soil biota, which increases the pathways by which nutrients such as nitrogen are assimilated by plants. Low nutrient conservation of intensive agriculture forces farmers to apply synthetic fertilizers to replenish lost nutrients. The nitrogen artificially fixed from the atmosphere to make fertilizers now exceeds biological nitrogen fixation by all terrestrial ecosystems. The regular application of nitrogen fertilizers to crops poses a threat to soil, water, and air quality due to nitrogen leaching and emissions of nitrous oxide (N_2O), a potent greenhouse gas that also degrades the ozone layer. In applying ecological knowledge, by recognizing the value of species mixtures in agricultural systems, we can improve nutrient cycling and conservation efficiencies, and reduce synthetic fertilizer use and its negative environmental impacts.

Nitrogen cycle

Groups of organisms, and interactions between them, control biogeochemical fluxes, including that of the nitrogen cycle. Nitrogen, a key nutrient for plant growth, cycles through atmospheric, terrestrial, and marine ecosystems in various molecular forms (Figure 14). In the soil, the process starts with the fixation of atmospheric nitrogen (N_2) by bacteria, or the conversion of organic nitrogen from the wastes and dead bodies of plants and animals into ammonium (NH_4^+), nitrite (NO_2^-), and nitrate (NO_2^-). Ammonium and nitrates are assimilated by plants, and transferred to animals that feed on them. Nitrogen is returned to the soil as wastes or dead bodies. Some nitrogen is lost back to the atmosphere from the soil through bacterial denitrification, the reduction process of nitrates to nitrogen gas (N_2), which completes the nitrogen cycle.

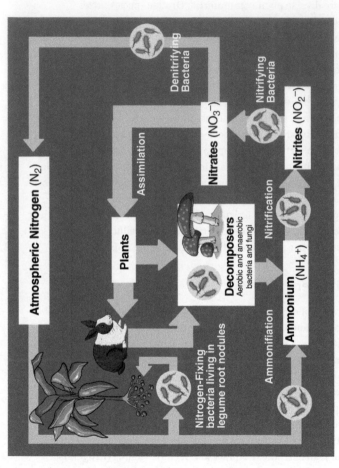

14. The ecology of mutualistic, competitive, and consumptive interactions between animals, plants, fungi, and bacteria plays a substantial role in the nitrogen cycle.

Plants are active players in this process. By co-opting soil fungi as mycorrhizal partners, and by forming other root symbioses with soil bacteria, plants accelerate the uptake of nitrogenous compounds and their conversion into plant biomass. Most of the 18,000 or so leguminous plant species, including such familiar species as beans, peas, and clover, form symbioses with nitrogen-fixing *Rhizobia* bacteria housed within root nodules. The plants supply the *Rhizobia* bacteria with carbohydrates, and in exchange receive nitrates that bacteria derive from gaseous atmospheric nitrogen. Alfalfa can fix over 200 kg of nitrogen per hectare in a year, while clover species fix around 150 kg per hectare annually.

Plant roots release carbon-rich exudates that stimulate bacterial conversion of soil organic matter to available forms of nitrogen, benefiting both microbes and plants. Some non-symbiotic free-living *Metarhizium* fungi provide as much as 48 per cent of plants' nitrogen requirements from insect tissue by infecting and consuming soil insects, with the plants releasing carbon from the roots in return.

Nitrogen losses from ecosystems occur through the leaching of nitrate into groundwater, or its conversion to nitrogen gas through microbial denitrification. Retention or loss of nitrogen is influenced by plant growth strategies and traits, and interactions between plants and nitrogen-cycling microbes. Rapidly growing plants reduce nitrogen losses by competing with bacteria for nitrates, thereby reducing the opportunity for nitrogen compounds to be lost from the ecosystem through leaching or microbial emissions. Some tropical *Brachiara* grasslands, and at least two temperate forage crop plants, alfalfa (*Medicago sativa*) and cocksfoot grass (*Dactylis glomerata*), reduce nitrogen leaching from soils by releasing chemicals that inhibit nitrifying micro-organisms, reducing ammonification and nitrification rates by 90 per cent. While this might seem counterproductive to plants that rely on the nitrates produced by these bacteria, these inhibitors are released only when ammonium concentrations in plant roots are high. Harnessing this conditional inhibitor of microbial activity has the potential to increase nitrogen use efficiency in agricultural systems. This would reduce the need for fertilizer inputs, and thereby avoid excessive nitrogen leaching into watercourses which could cause algal blooms, to the detriment of aquatic insects and fish.

While different plant species vary in their capacities to influence nitrogen-cycling processes, subject to their resource acquisition strategies and associated traits, the ecology of plant–fungi–bacteria interactions has major implications for ecosystem fluxes of nitrogen and other nutrients, and determines the nature of the nitrogen cycle at larger scales.

Back to natural history

Raymond Lindeman's great insight was to interpret complex ecosystems as flows of energy and matter between biotic and abiotic ecosystem components. He is remembered for his quantitative and conceptually novel study of an entire ecosystem, including its micro-organisms, plants, animals, and non-living components. Less well known is that the young Raymond spent his childhood fully engaged in his love for natural history. His ecosystems understanding, encompassing the concept of food webs and energy transfers, owes much to his amblings at the rough margins of the family farm.

Chapter 5
Simple complex questions

In ecology, simple questions have complex answers. The most basic questions, as Paul Colinvaux noted four decades ago, begin with 'why'. Why is the world green? Why are there so many species? Why are big fierce animals rare? Why should we care about biodiversity? These deceptively simple questions invite us to delve into the fundamental principles of ecological theory. The 'how' questions are about the mechanics of ecology, the processes by which populations, communities, and ecosystems work. The 'how' questions relate process to pattern, enabling us to manage ecosystems appropriately, but to uncover ecology's generalities and laws we must ponder the 'why' questions.

Why is the world green?

Pity the poor farmer's constant toil against insect pests and fungal pathogens that threaten his crops. He deploys an arsenal of chemical pesticides to keep these enemies at bay, and to keep his fields green and productive. Our food production systems depend on our ingenuity in developing new poisons to protect our crops. If we lower our guard, we risk being overwhelmed by all manner of herbivores that threaten to strip our crops like a veritable plague of locusts. The agricultural industry is predicated on the simple ecological observation that an abundance of resources (i.e. crops) is a boon to consumers (herbivores), unless such consumers are kept in check by predators or pesticides.

Such thinking fails to explain why most of the land surface of the planet, from the tropics to the cold temperate boreal zones, is luxuriously green. This riot of vegetation, to paraphrase Charles Darwin, begs an explanation. What of the pests that threaten humankind's agricultural endeavours? Why do they not strip away Earth's abundance of leafy biomass, as they so readily do our agricultural fields? Why is the world green?

This simple question forms the title of a classic 1960 paper (by Nelson Hairston and co-authors) that concludes the world is green because herbivores and other plant pests are kept low by predators, pathogens, and parasites. Predators control the abundances of herbivores, which alleviates consumer pressure on plants. Yet life, and ecology, is rather more complicated than first meets the eye. For one thing, the corollary of these 'top-down effects' is 'bottom-up effects', where plants control the abundances of consumers higher up the food chain. This implies that despite seemingly abundant plant food for herbivores, plants somehow limit access to this food, and so limit herbivore numbers. Predators, in turn, are restricted by the abundances of their prey. In this scenario, plants are firmly in control. Both top-down and bottom-up theories are plausible, and we must evaluate the evidence for both to understand why the world is green.

Top-down control

To determine the extent to which predators control natural systems we simply need to remove predators, and wait to see what happens. This is, all too tragically, an ongoing unplanned experiment, achieved through our thoughtless extirpation of many large predators from many of the world's biotas. We can, at least, draw some lessons from this sad history.

Our expectation, assuming top-down control, is that herbivores released from predation should proliferate and eat up all the vegetation. Red deer in Scotland roam the land in abundance, free of predators as wolves and lynx have long been extinct in the UK. Abundant deer browse young tree saplings and prevent forest regeneration in all but the most inaccessible places. The hills of the Scottish Highlands remain mostly bare of trees, retaining only a thin covering of grasses, heather, or bracken fern. Efforts to establish new woodlands in the Highlands rely on fencing off large areas to prevent access to deer, or intensive culling of deer populations. This has resulted in substantial woodland recovery and a greening of the land. The increase in plant biomass following simulated predation (in the form of culling) suggests the world is green due to top-down control by predators on herbivores.

The small forested islands created by the flooding of Lago Guri reservoir in Venezuela, and studied by John Terborgh, are not large enough to support viable populations of armadillos and primates. Both these mammals are predators of leaf-cutter ants. Released from predation, leaf-cutter ants increased dramatically, leading to marked reduction in tree saplings, and the proliferation of ant-resistant lianas and vines. Vine-covered islands are still green. The small islands did not experience a change in plant biomass, but rather a shift in species composition towards less palatable plants. These subsequently limited the ability of ants to sustain large populations, even in the absence of predators, implying the action of bottom-up effects as plants control herbivores. This outcome implies that both top-down and bottom-up effects play a role in structuring ecological communities.

The importance of top-down control is influenced by factors that affect predator efficiencies, including climate. On Isle Royale in Lake Superior, wolves form larger packs in particularly severe winters. This increases their success in hunting moose. Reduced browsing by moose on balsam fir (*Abies balsamea*) saplings allows rapid sapling growth, as revealed by wider tree rings. In milder winters, wolf packs are smaller and wolves seek alternative prey, and fir tree saplings experience slower growth and higher mortality, which allows other plant species to spread. Climatic fluctuations can alter ecosystems by shaping the behaviour of predators, which cascades down to affect plant communities.

Green deserts

What of bottom-up processes? How do plants limit herbivore numbers? Many plants use physical defences to protect against herbivory—consider the spiny thorns of cacti and acacias, the stinging nettles (*Urtica dioica*) familiar in Europe, or the related but altogether more fearsomely stinging gympie-gympie tree (*Dendrocnide moroides*) of Queensland (Figure 15). Leaf blades of grasses contain silica crystals that reduce digestibility and wear down insect mandibles and mammalian teeth. Entrapped sand on the sticky surfaces of sand verbenas (*Abronia* species), native to the west coast of North America, performs a similar function, and deters caterpillars from feeding on them.

Other plants pack their leaves and stems with an array of noxious chemicals. The astringent and spicy flavours we enjoy in our cuisines are derived from plant chemicals that are, essentially, toxic to potential herbivores. Human history and literature is replete with references to poisonous plants. Socrates drank hemlock (*Conium maculatum*), which is rich in deadly piperidine alkaloids. Macbeth's soldiers poisoned invading Danes with wine made from the sweet fruit of deadly nightshade (*Atropa belladonna*). Castor beans contain ricin, deadly in even small amounts, which was weaponized by both the United States and Soviet Union.

Plant toxins are secondary metabolites, compounds that are not involved in plant growth, development, or reproduction, but instead protect plants from herbivory and microbial infection.

15. The gympie-gympie, or stinging tree, of Queensland, Australia, packs a very painful and debilitating sting sufficient to deter mammalian herbivores (and people who learn from experience to give it a wide berth). Despite this, at least some insects appear to be immune from these stings, and readily munch on the leaves.

Australian *Eucalyptus* trees produce monoterpenes that are potent deterrents to brushtail possums, creosote bushes in the western United States have phenolic resins that limit consumption by desert woodrats, and birch trees in boreal North America use papyriferic acid to deter snowshoe hares. Plants also use a range of alkaloids, phenolics, and tannins to protect their leaves from insect herbivores. While some secondary defence chemicals are downright poisonous, others simply make plant tissue unpalatable or distasteful, and herbivores learn to avoid them. Toxic compounds reduce herbivory, but are costly in the resources required to produce them. Plants that have high concentrations of secondary compounds tend to be slower growing than chemically undefended plants.

Whether physically or chemically defended, plants substantially reduce what herbivores eat, or at least reduce the rate at which they consume plants. What appears to be a bountiful landscape is, to prospective herbivores, more akin to a green desert. Plants also have low nutritional quality relative to what animals need. Animal tissue is around ten times as rich in nitrogen as plant tissue, and so herbivores must acquire nitrogen from plants that are comparatively deficient in the nutrient they need. Herbivores must consume much more plant material than is necessary for energy provision alone simply to secure their nitrogen needs. This large demand for plant biomass implies that herbivores should quickly turn the world brown, but this is clearly not the case. Instead, the time and energy spent consuming plant tissue exposes herbivores to predators, and leaves less time for reproduction. Both factors keep herbivore populations low.

If this theory is correct, then increasing the nutrient content of plants by adding fertilizer should support higher herbivore biomass. Fertilizers applied to heathlands in the Netherlands have indeed increased the number of heather beetles, which has reduced heather dominance and allowed purple moor grass to spread through the plant community. As with top-down control, this outcome does not so much reduce overall plant biomass, but rather shifts plant community composition towards less palatable species.

Specialists

Herbivores have responded to plant defences by evolving mechanisms to detoxify or circumvent plant poisons. Given the variety of plant toxins, herbivores are not able to evolve defences to all, and so evolved responses tend to cause herbivores to specialize on particular plant groups, or even single species. Caterpillars of cinnabar moths (*Tyria jacobaeae*) feed exclusively on ragwort (*Jacobaea vulgaris*) that debilitates, or even kills, horses and cows. The cinnabar caterpillar sequesters the ragwort

16. Specialist herbivores such as the cinnabar moth caterpillar have evolved abilities to overcome plant defences, and thereby gain exclusive access to an otherwise poisonous plant resource.

toxins to protect itself against predation by birds, and advertises its unpalatability by its bright yellow and black banding (Figure 16). Vertebrate herbivores have similarly evolved abilities to deal with toxins in particular plant groups, and so become restricted to these groups. Koalas in Australia feed only on the monoterpene-rich eucalyptus trees that few other animals can handle.

Plant toxins explain why most herbivores cannot eat most plants, but why do specialist herbivores not consume all their food supply? For specialist vertebrate herbivores, toxins set upper limits on food intake. Herbivores must avoid saturating their detoxification systems, which occurs before ingestion rates are limited by mechanical handling of food items. Koalas, for example, eat much less foliage if forced to feed on the better-protected eucalyptus species.

If the majority of plant species are toxic or indigestible to all but specialist herbivores, how is it that humans have such catholic eating habits? In truth, many of the plants we eat *are* toxic, some highly so in their original wild forms, but through millennia of plant breeding we have selected for the most palatable varieties. Even familiar vegetables, including potatoes and tomatoes of the Solanaceae family, are highly toxic in their original wild forms. Our close relatives, by contrast, are highly selective in their choice of plant food. Mountain gorillas eat a very small percentage of the total number of plants in the forest, and avoid some of the most abundant plants altogether.

So why is the world green? The Earth's land surface greenness is due to a combination of bottom-up and top-down processes, as well as climate, and other environmental factors, all of which collectively control rates of herbivory. The dual role of these processes is apparent in tropical forests where predators limit the abundance of insect herbivores feeding on rapidly growing plants in clearings, while in the deep shade of the forest understorey bottom-up processes dominate as slow growing plants invest in leaf defences that limit herbivore activity and numbers. Overall, top-down processes might be less important than bottom-up processes in maintaining a green world, if only because they do not so much determine plant biomass, but rather the particular composition of plants. Bottom-up processes are the most widely applicable explanation for why herbivores do not destroy all vegetation, so much so that Coleridge's 'water, water everywhere, nor any drop to drink' might for the terrestrial realm be adapted to 'food, food everywhere, nor a bite to eat'.

Why so many species?

Scientists at Lambir Hills, a forest reserve lying just north of the equator in Sarawak, Malaysian Borneo, have recorded 1,008 tree species in a single 50-hectare plot. The entire tree flora of the

United States and Canada, by comparison, comprises only around 700 species. We find the same pattern of exceptionally high tropical richness in many animal groups, including reptiles, fish, birds, mammals, and invertebrates. Species numbers decline towards temperate regions, and even more so in boreal zones. Why so many tropical species? Why relatively few species outside the tropics?

The answer to global patterns of species distributions takes us beyond the realm of ecology and into the disciplines of biogeography and evolution. It is likely to have something to do with climate, and climatic unpredictability, acting on evolutionary time scales over which speciation and extinction unfold. Some of these explanations infer climatic instabilities over geological time spans that increase extinction rates and limit possibilities for speciation. The warmer temperatures and smaller climatic variations of the tropics provide a basis for an interesting evolutionary-ecological explanation of tropical richness. Dan Janzen argued that high mountain passes are more insurmountable barriers to tropical organisms that are not adapted to cool temperatures. Species of higher latitudes, on the other hand, are annually exposed to cold temperatures and have wider physiological temperature tolerances, and so have less difficulty in crossing mountain passes. Janzen used this inference to argue that tropical species have smaller ranges constrained by their physiological temperature tolerances, which creates opportunities for speciation through isolation and divergence of populations.

Ecological explanations of species richness focus on the mechanisms that allow species to coexist together, as they are more likely to do in high diversity areas. This brings us back to the question of how a tropical forest can support over 1,000 tree species in a 50-hectare plot. This is not to say that temperate habitats are uniformly species poor. European calcareous grasslands can include over forty species of herbs and grasses within a single square metre of meadow. The ecologically relevant question is how are so many species (be they tropical forest trees or temperate meadow herbs) able to coexist given that Gause's competitive exclusion principle prohibits this.

A solution to this conundrum is that species avoid competition through specialization, in much the same way that MacArthur's warblers foraged for insects on different parts of the same tree. Competition is avoided through specialization across environmental and resource gradients. A habitat is able to support many more specialist species, each with narrow and minimally overlapping requirements, than generalist species with broadly overlapping niches.

Coexistence through specialization raises the question as to whether there are sufficient numbers of niches to support the multitude of species occurring in any one place. Collectively, species have a bewildering array of traits that underpin growth, survival, and reproduction. Tree species in a forest, for example, differ in their growth and survival responses to light conditions, soil water and nutrient availabilities, herbivore pressure, and disturbance events. Trees also have different regeneration strategies, some producing few large seeds, others opting for large numbers of small seeds. Trade-offs among strategies and their associated traits prevents any single species from dominating all environmental conditions. Multiple trade-offs create a multiplicity of strategies among species, while environmental and biotic variability creates an array of opportunities to which these strategies can be applied. A falling tree creates a gap in the canopy of the forest that alters the light and soil environments, the depth of the leaf litter, the interception and infiltration of rainfall, the microclimate, and the biota, from the canopy to the forest floor, and from the centre of the gap to its edges. A forest canopy gap therefore encompasses many microhabitats, each subtly different according to variation in light, soil, and biota. Seedlings of many different tree species establish initially in these forest gap microhabitats by chance, but are soon sorted through differential growth performance based on the extent to which their traits suit the local light, soil, and microclimatic conditions, and competitive interactions. A diversity of microhabitats sets the stage for a sorting of species according to their adaptive traits and trade-offs.

In view of the many environmental and biotic gradients and combination possibilities, there are a large number of potential niches, and hence opportunities for coexistence of many specialist species. In practice, it is often difficult to directly link the distribution of forest trees, based on their trait combinations, to such fine-scale environmental complexity. Local environmental conditions can reflect tree species performance and survival likelihood, but with relatively little precision. Fine-scale niche differentiation is likely to be only a partial explanation of high numbers of coexisting species.

Density dependence

Just as the competitive exclusion principle is a central tenet of ecology, so is density dependence, and it should come as little surprise that density dependence mechanisms have been co-opted to explain tropical species richness. In the early 1970s, Dan Janzen and Joseph Connell, working independently, argued that high numbers of specialist insect herbivores, or fungal pathogens, could determine the survival of tree seedlings in a density-dependent manner, thereby maintaining species diversity. Most seeds do not disperse far from the parent tree, where seedlings are found at high densities, and at much lower densities further away. Carpets of seedlings close to the mother tree are vulnerable to attack by specialist herbivores and pathogens that readily move from the mother tree to nearby seedlings (Figure 17). Seedlings therefore struggle to establish and mature close to their parents or other trees of the same species. Long-distance dispersal allows seedlings

17. A dense carpet of seedlings of *Shorea gibbosa*, in Borneo, is vulnerable to attack from herbivores and seedling pathogens and few if any of these seedlings are likely to survive, whereas a more distantly dispersed and isolated seedling is likely to escape pest and pathogen attack.

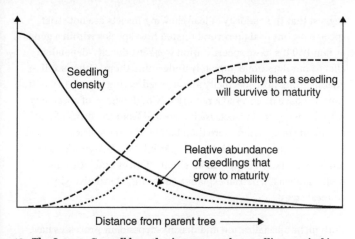

Distance from parent tree ⟶

18. The Janzen–Connell hypothesis proposes that seedling survival is highest at intermediate distances from parent trees. Most seeds fall close to the mother tree, but here they are exposed to very high herbivory and pathogen pressure that destroy almost all seeds and seedlings. Survival is greatest at intermediate distances which still receive some seedfall, but where pathogens and herbivore pressures are much less.

to escape pests and pathogens (Figure 18). Well dispersed seeds establish amongst the seedlings of other tree species, where their distance from the mother tree and their low densities give them a far better chance of escaping notice by specialist herbivores or pathogenic fungi. This encourages a mix of different species in a local area, and tends to suppress the formation of high-density aggregations of seedlings of the same species.

A key assumption of Janzen–Connell theory is that tropical herbivores (and pathogens) are specialists, whereas in temperate communities they are less so. If generalist herbivores prevail, then seed dispersal and low density would offer no advantage given that seedlings would be vulnerable to attack from any generalist herbivore in neighbouring trees. Painstaking studies of herbivores and their host plants in New Guinea's remote tropical forests suggest that the majority of leaf-chewing insects are not strict specialists, but feed on several related host species within a genus or family. This loose specialization weakens density-dependent effects, but does not completely undermine them. Other studies documenting host specificity among seed-eating insects in Central America have discovered a remarkably high degree of specificity, with 80 per cent of insects each recorded from the fruits of just one plant species, and more than half of the tree species attacked by no more than two insect species. While the Janzen–Connell model is unlikely to apply in all circumstances and locations, it offers at least a partial explanation for high local diversity of tropical trees.

Both niche specialization and density-dependent processes have some degree of support as explanations for species coexistence. They are not mutually exclusive, and are likely to operate together alongside other drivers of high species richness such as energy-productivity processes or disturbance. As in much of ecology, there is a plurality of ecological and evolutionary mechanisms at play.

Biodiversity: what is it good for?

Early in 2019 the Intergovernmental Science-Policy Platform on Biodiversity and Ecosystem Services (IPBES) published a report bemoaning substantial and ongoing losses of species across the globe, expressed as declining populations, the loss of whole populations from many areas, and, in the worst cases, global extinctions of species. The report estimated current extinction rates to be three orders of magnitude higher than might be expected in the absence of humans. Yet there are over 60,000 species of tree in the world, and 391,000 vascular plants. The number of documented insect species is 925,000, though estimates put the total number at around five million. Fungi too likely number more than five million. Do we need so many species? Is there is any ecological reason to justify concerns about declining biodiversity, life's richness and variety?

A first step in answering this question is to figure out what species do. Ecosystem functions are the effects of biota on the biological, physical, and chemical properties of the environment, including the fluxes of energy, nutrients, and materials through environments. Associated with this are ecosystem services, the natural processes that contribute to human wellbeing through, for example, food production by pollination, nutrient cycling in soils, diminished flood risks by rainfall interception and slowed water flows, mitigation of climate change by plants that sequester atmospheric carbon dioxide, and improved physical and mental health through pleasant environments for recreation and relaxation. Ecosystem functions and services arise from interactions among species and their environment. If biodiversity determines ecosystem functions and services, then demonstrating this relationship, and linking biodiversity to tangible human benefits, provides a powerful argument for biodiversity conservation.

Given the complexity of natural communities, and the sheer number of species involved, ecologists assign species with similar traits and strategies to functional groups. Among plants, functional groups include nitrogen-fixers, annual weeds, or evergreen shrubs. In tundra ecosystems, most vascular plants can be allocated to four functional groups: evergreen shrubs, deciduous shrubs, graminoids (grasses and sedges), and forbs. In tropical systems there are many additional plant functional groups, including fast growing early successional trees, competitive canopy trees, and lianas. Herbivore functional groups include migratory grazers (many ungulates), sedentary grazers (leaf-chewing insects), browsers (deer or giraffes), wood feeders (termites and elephants), root feeders (insects and mammals), and a variety of insect stem borers, leaf miners, gall-formers, or sap-suckers. These groups each contribute to ecosystem functions in different ways.

Herbivores and decomposers contribute to nutrient cycling, while pollinators and seed dispersers promote plant reproduction.

Functional redundancy

Charles Darwin noted that a mixture of grass species in a meadow produces more herbage than a single species growing alone. Variation in rooting depths among species allows a greater range of soil depths to be exploited. This niche complementarity, whereby different species utilize different parts of a resource in a complex environment, improves the efficiency of resource acquisition across the functional group, thereby enhancing ecosystem function. Despite niche complementarities, species within functional groups overlap in their functional contributions. One species is therefore substitutable, at least to some degree, by another species in the same group. Niche complementarity argues for the need for species richness to enhance ecosystem function, while the substitutability or redundancy of species indicates otherwise. We need to know how biodiversity affects ecosystem functions in the context of complementarities and redundancies. A first approximation might assume that the provision of ecosystem functions improves with every new added species, each of which contributes in its own unique way. In fact, the redundancy among species means that while functional performance improves as each added species introduces complementarity, marginal benefits begin to level off with increasing redundancies among species until no further benefit is gained. If we were to reverse this process, redundancies among species in a species-rich community implies that we can, initially at least, lose species without incurring much loss of function.

Redundancy within functional groups provides insurance. The loss of some species can be offset by an increase in the activities of others in the same functional group. A plethora of species also increases the likelihood that at least some will be tolerant of any perturbations that might afflict the community. A greater number of species is also more likely, by chance alone, to include particularly productive, or particularly resilient, species that will continue to deliver ecosystem functions in the face of external disturbances. Of greater ecological interest is that species using the same resource are likely to differ in the environmental conditions in which they best perform, which facilitates complementarity in a seasonal or changing environment.

Ecologists have been testing these ideas with experiments, and an ecologist's principal tool is the field experiment. Numerous field experiments have sought to test the theory that species diversity increases ecosystem function, the latter usually measured as increasing community biomass. These experiments establish different mixtures of plant species arranged in species-poor and species-rich combinations (Figure 19). After a time, the productivity of these blocks is measured. Blocks with the greatest number of species are expected to gain most biomass, for reasons of niche complementarity, and the chance effect of including a particularly productive species. Most of these experiments show

19. **Experimental plots of different species mixtures established in 1994 as the 'Big Biodiversity' experiment at the University of Minnesota to evaluate relationships between species diversity and ecosystem function.**

that productivity does indeed increase as the diversity of species increases. Yet this relationship tails off relatively quickly, suggesting gradually diminishing returns in productivity gain with each additional species. This continues until gains drop to zero, so that any further species added to the mixture is functionally redundant. Experiments indicate that this saturation of ecosystem function occurs with relatively few species, and far fewer than occur in natural systems. Nature is, it seems, rich in redundant species.

Most experiments focus on productivity and biomass increase. We assume that the performance of many other ecosystem processes is improved by increasing species numbers. Recycling of nutrients, for example, benefits from different species playing different functional roles, by tearing apart large woody debris, shredding leaf material, chewing lignin and cellulose, chemically dissolving and digesting plant fragments, and bioturbation of the soil that redistributes organic matter. Chewers, shredders, digesters, and redistributors are functional groups, each with many species representatives. In each case though, there appear to be far more species than is necessary to deliver a fully functional nutrient cycling service.

Redundancy begets stability

For humankind, substantive functional redundancy comes as something of a relief given our impacts on biodiversity. Yet the species pool is never static. Species are more or less successful given changing conditions and the ebb and flow of resources, seasons, and disturbances. In any single functional group, large numbers of species provide a buffer or insurance against environmental fluctuations that affect some of these species, but are unlikely to affect all. We expect redundancy to deliver stability.

A variety of species allows the community to thrive even when subjected to changing environmental conditions and disturbances.

Diverse communities are better able to resist invasive species, as putative invasives struggle to establish in species-rich communities where available resources are already finely divided and captured by native species. The broader range of attributes in a species-rich community provides community-wide adaptive capacity to environmental stresses. A portfolio of species is similar to a financial portfolio that spreads risk across a variety of investment types and mechanisms. Diversity provides resilience against change.

Stability in complex systems

Species-rich ecosystems have a bewilderingly complex range of feeding interactions. Even a simplified food web of the North Atlantic marine ecosystem is a mess of incoherent links (Figure 20(a)). A closer look reveals a few major channels through which the large share of energy flows, from plants and photosynthetic plankton as the primary producers, through to the top predators. We can, rather glibly, reduce our complex North Atlantic food web to these main channels, and redraw it as seals, which feed on cod, both of which feed on 'everything else' (Figure 20(b)). In terms of energy and nutrient flows, this simplified food web fairly captures the essence of the otherwise complex North Atlantic food web. The implication is that there is substantial redundancy in food webs if our interest is to understand large-scale flows of energy and resources. Ignoring the food web's minor links and peripheral species makes modelling of the productivity of fisheries much simpler.

Yet it turns out that the minor channels and peripheral species that we might be keen to disregard play an important role in stabilizing the food web. The many weakly interacting species collectively counter fluctuations in populations of the dominant species. Without these minor species, a simple food web would be subject to wild fluctuations and oscillations akin to those of lemmings on the Arctic tundra. By appropriating a share of the

(a)

(b)

seals

cod

everything else

20. Two representations of the North Atlantic Ocean food web, one simplified (a), the other very simplified (b).

resources, the weak interactors dampen fluctuations in resource availability or consumption, and hence stabilize the food web and the community as a whole.

There is an important lesson in this for us. It is not sufficient to confine our attention to the dominant species when managing natural resources and ecosystems. We also need to heed the losses of the rarer and less obvious species. These, collectively, play important roles in buffering ecosystems from external perturbations and internal fluctuations, while contributing through complementarities and redundancies to the continued provisioning of ecosystem functions and services upon which humanity depends.

Ecologists are often asked how much biodiversity humanity actually needs. Such a question implies a solely human-orientated utilitarian perspective on nature that most ecologists, and perhaps most people, would object to. Setting this issue aside for a moment, the answer is that there is no precise answer, but ecological science suggests the more biodiversity the better. Earth's biological richness does, of course, have value well beyond its function for human wellbeing. Environmental policies and actions should, therefore, also respond to a conservation ethic and not only to functional benefits.

Chapter 6
Applied ecology

Our greatest challenges are primarily environmental. Climate change, biodiversity loss, land degradation, pollution and plastics, nitrogen deposition, and invasive species provide a litany of issues that we as a global society need to address now and for decades to come. The scale of these problems is compounded by the growing size and wealth of the human population. We have long been warned of these issues. Aldo Leopold and Rachel Carson, among many others, counselled us to embed ecological principles into the management of land and our socio-political systems. By and large, we have not done so. There are many reasons why we have not, but a small part of the explanation is that our ecological knowledge has not been sufficiently well developed to provide the conceptual and methodological tools for appropriate management guidance. A much larger part of the reason for abject environmental failure lies with the operations of dominant global socio-economic and political systems, a topic for another book.

Applying ecology

Ecology underlies many of the principles, concepts, theories, models, and methods to address environmental problems. Applied ecology strives for practical environmental solutions by developing management options based on ecological theory, by projecting plausible trajectories of change, and by evaluating outcomes. We apply ecological knowledge to model renewable resources, such as fish or timber, to provide guidance on how to exploit them without diminishing long-term productivity. Applied ecology also deals with the species we wish to control, including agricultural pests and weeds, invasive species, and animal and plant diseases. Equally, we use ecological theories and methods to protect species we value, through either conservation action or habitat management.

The natural world provides a variety of services that benefit humanity, including pollination, hydrological regulation, nutrient cycling, and carbon sequestration. Managing the environment to maintain ecosystem services extends the realm of applied ecology to broader landscape scales. This encompasses the management of forests, agro-ecosystems, rangelands, peatlands, mountains, coastlines and seascapes, and more. In landscape settings we are forced to work with processes that extend across not only spatial scales, but also temporal scales, from short-term population fluctuations and acute disturbances, to longer-term shifts in soil fertility, habitat composition, or climate.

Applied ecologists work at the intersection of human actions and environmental outcomes, and by its very nature applied ecology is interdisciplinary. Ecological theory and principles suggest ecologically sound management but, in the end, management decisions are implemented by farmers, foresters, businesses, and policy-makers, all of whom have their own set of considerations, needs, and priorities. To understand environmental decision-making and planning, applied ecology has to rub shoulders with other, often substantially different, disciplines such as economics, policy, ethics, behavioural psychology, mathematics, and environmental law. The use and application of ecology in management is therefore rather messier than the scientific discipline of ecology might suggest. An ecological approach to management takes account of interactions among organisms and their environment, and considers feedbacks across space and time in complex landscapes, but, crucially, must also do this in the context of human-oriented norms and needs. This makes applied ecology in the broader context of decision-making probably the most complex and challenging of ecological disciplines, as well as the most needed.

Maximum sustained yield

Ecological theory is used to model population dynamics, by which sustainable harvests can be estimated. Fisheries scientists use models to make recommendations on quotas and fishing effort to ensure stable and viable fisheries. Ecological theory expounds that populations of wild species grow to their carrying capacities, defined as the maximum population size that can be supported by the environment, when the number of births equals deaths. Reducing the population alleviates density-dependent competition for resources, leading to population recovery and a return to the carrying capacity. A small reduction in the population reduces density only slightly, and the resulting surplus of births over deaths is marginal. A large population reduction alleviates competition substantially, but because the resulting population size is small, the number of births is concomitantly low. Somewhere between these two extremes is a population size sufficiently large to generate many new births, but low enough that alleviated competition means low mortality. Assuming only density-dependent regulation, the greatest difference between births and deaths occurs at a population size that is exactly half of the carrying capacity. At this population size where growth rate is highest, it should be possible to harvest this maximum surplus, the number of births minus deaths, in the expectation of no overall change in the population, which will continue to generate the surplus *ad infinitum*.

This harvest level is the 'maximum sustained yield'. It is tempting to manage renewable resources on this basis, and fisheries often have done so. Harvesting the maximum sustained yield supposes we have certainty about population size and density-dependent growth rates. If recommended harvest is set slightly higher than the maximum sustained yield, the individuals harvested will, year on year, exceed the surplus generated. The population will decrease, slowly at first, but then with increasing rapidity to extinction. If the offtake is somewhat less than the maximum sustained yield, the population will settle at a higher stable equilibrium, but will generate a lower yield.

All this makes the assumption that the harvested population is a closed system, not affected by anything other than internal density-dependent regulation. Ecological systems are, however, at any scale of analysis, dependent on and affected by processes that lie outside the scale of consideration. One such external phenomenon is the El Niño Southern Oscillation, a recurrent climatic event that warms sea surface temperatures and alters patterns of oceanic upwelling, causing greatly depleted nutrients in coastal waters off South America. An El Niño in 1972 caused the collapse of the anchoveta fishery, which remained low until the 1990s. Continued fishing at the maximum sustained yield, based on a model that did not consider this fluctuation, compounded the slow recovery of the anchoveta stock.

Cod

Even the assumption of an equilibrium carrying capacity regulated by density dependence is often not valid. Such an approach considers one species at a time, without incorporating the interactions among species in the community.

Consider cod. In 1497, fish 'so thick by the shore that we hardly have been able to row a boat through them' seriously impeded John Cabot's progress along the coast of Newfoundland. Baskets lowered into the water by the side of the vessel could be retrieved moments later full of fish. Some 400 years later Thomas Henry Huxley, giving the inaugural address at the 1883 Fisheries Exhibition in London, stated with confidence, 'I believe, then, that the cod fishery, the herring fishery, the pilchard fishery, the mackerel fishery, and probably all the great sea fisheries, are inexhaustible.' A hundred years later still, in 1992, the great cod fisheries of the Grand Banks had collapsed. What went wrong?

Cod feed on smaller fish, squid, and crabs, which in turn feed on zooplankton that feed on phytoplankton. Changes in the abundance of phytoplankton, the productivity of which is subject to nutrient upwelling from ocean depths, ultimately affect cod near the top of the food web. Additional feedbacks complicate matters further, as juvenile cod are eaten by fish that are themselves the prey of adult cod. Harvesting adult cod releases these 'meso-predators' from predation pressure, which imposes more predation on the juvenile cod. Meso-predators become top predators, and by eating juvenile cod they prevent the recovery of cod stocks. Cod persist in the Northwest Atlantic, but not in sufficient numbers to support a large fishery, and populations have shown no signs of recovery since the 1992 collapse. The Northwest Atlantic marine ecosystem can be said to have flipped into an alternative state, dominated by meso-predators that suppress juvenile cod numbers and prevent recovery. The food web, now restructured into a new self-sustaining state, precludes the re-establishment of adult cod at the top of the food chain.

Estimating the size of fish stocks is bound to be challenging, but ecologists now have far more sophisticated monitoring and modelling systems. Crucially, ecologists also recognize that a closed model based on density-dependent processes is not sufficient to account for population fluctuations. New models incorporate food web interactions, and estimate populations of both prey and predator species, as well as the species of harvest interest. They take account of climatic changes and weather patterns that affect nutrients. Fisheries scientists also quantify feedbacks that might risk shifts to undesirable states. Improvements in high speed computing facilitate the analysis of these data.

Advances in multispecies analysis are, inevitably, more demanding of quantitative information to feed the models. Yet it is not only the fallibility of models that is of concern for the viable management of fisheries, but also the willingness of politicians and the fishing industry to accept and abide by the recommendations of fisheries ecologists, and to accurately report catches. Fishing regulations set quotas on the amount of harvestable fish, or limit fishing effort by restricting boat numbers, fishing days, or the type of equipment used. Such restrictions are politically contentious as they reduce the potential earnings of the fishing industry, and affect livelihoods in fishing communities. They are also difficult to enforce. Substantial quantities of caught fish, including immature cod, are dumped back into the sea as they have low market value or are non-target species for which boats have no licence. This bycatch, coupled with misreporting of catches, reduces the reliability of fisheries models. The future management of fisheries cannot therefore rely solely on ecological theory and models, but must couple ecology with a social understanding of the behavioural responses of policy makers and fishing communities.

Invasive species

Alien invasive species introduced, misguidedly or inadvertently, into geographies beyond their ordinary range have caused immense environmental and economic damage, especially on islands. Owing to their isolation, islands are generally species poor, but comparatively rich in endemic species that are not found elsewhere. These endemics, protected from the competitive rigours of much larger biotic communities on continents, are especially prone to invasive species. Almost 60 per cent of modern species extinctions have occurred on islands, in large part due to invasive species impacts.

Charles Elton recognized the risks of alien species in his 1958 book *The Ecology of Invasions by Animals and Plants*, in which he described the extent and manner by which humans have promoted animal and plant invasions. Elton sought to understand the ecological factors that facilitate or impede biological invasions at various stages in the invasion process. The success of many alien species might be predictable by what we know of the regulatory pressures imposed by predators, competitors, parasites, and diseases to which these species are exposed in their native range. Where there are equivalent species in the native community that feed on or compete with introduced species, the spread and impact of the alien is likely to be limited. It is often the case, however, that alien species are advantaged in their introduced range as they are no longer exposed to the predators and competitors of their native range. Release from these enemies allows them to spread rapidly, often at the expense of native species.

Early efforts at controlling invasive species on islands often end disastrously owing to insufficient attention to basic ecology. Invasive species have evolved alongside their predators, so introducing a predator might reduce an invasive, but it is unlikely to eliminate it. An introduced predator might more likely find native species that have no prior experience of the introduced predator an easier catch. This renders a naive native biota very vulnerable to extinction. Releases of cats, dogs, or mongooses to control invasive rats have thus had modest impacts on the target rat populations, but they have devastated many native species. Endeavours in the 19th century to control rodents in Jamaican sugar cane fields left a litany of disasters. First, the European red wood ant was brought in to discourage rats, and not only failed to do so, but soon became a problem in itself. Cane toads introduced from mainland America to remove both rats and ants also became pests, while ants and rats persisted. Finally, farmers turned to the Indian mongoose to control rats and toads. Mongooses discovered that the native bird population was easier prey, creating a new set of problems.

While we now have better ecological knowledge of predator–prey dynamics, we do not always use it effectively. Control programmes often lack ecosystem perspectives that consider the trophic and competitive interactions among species in communities, and we continue to be surprised by outcomes. The successful removal of introduced goats and pigs in 2000 from a threatened native forest on Sarigan Island in the tropical western Pacific Ocean had the unfortunate but potentially predictable consequence of releasing an invasive vine, *Operculina ventricosa*, from grazing pressure, resulting in its spread through the native forest. Similarly, dramatic explosions of formerly overlooked introduced mice have followed rat eradication. Removal of an introduced predator, competitor, or herbivore can result in the increase of another previously suppressed alien species, to the ongoing cost of native species.

Eradication programmes drawing on good knowledge of species interactions across the community have more success. The removal of invasive black rats in 2006 from Surprise Island in the Entrecasteaux Reef, New Caledonia, was preceded by a four-year study of the island's flora and fauna to identify the impact of rats on native species, and to reveal the presence of other introduced species such as mice, ants, and plants. The study included plant and vertebrate surveys, diet analyses, food web characterization, and population dynamics modelling. Modelling trophic relationships in the invaded ecosystem indicated that removal of the rats alone would trigger the release of a small alien mouse population. Both rodent species were therefore removed, and unfortunate surprises on Surprise Island have so far been avoided.

Controlling invasive species on mainland populations is more challenging, as complete eradication is almost impossible. Occasional long-distance dispersal can also spark new outbreak sites far from the original range. Invading organisms at the expansion frontier even evolve improved dispersal capabilities,

Box 3 Cane toad

Sugar cane farmers in 1930s Queensland, Australia, had a problem. Cane beetles were destroying their crops. Someone told somebody else of a toad that had a voracious appetite for the destructive beetles. In 1935, over 100 toads travelled in two suitcases from Hawaii to Australia, to be introduced into cane fields around Cairns and Gordonvale. They ignored the beetles, wandered off, and multiplied. They now number more than 1.5 billion, extending over 1 million square kilometres of Queensland and the Northern Territories. They have eaten their way through much of Australia's native plants and animals. Being toxic, they kill any would-be predators. There is little to stop their continued expansion across all but the driest of Australian landscapes.

which allows them to more rapidly expand into new areas. Cane toads (Box 3) spreading westwards across northern Australia have evolved longer hind legs and a tendency to continue moving in a straight line, which has increased their rate of spread from around 5 km per year when first introduced to Australia, to the current 50 km per year. This accelerates invasion rates, and increases the scale of the problem. In such situations, we simply have to learn how to live with the invasion.

Pest management

Globally, sales of pesticides to protect agricultural crops are estimated at $52 billion for 2019. Every dollar spent on pesticides leverages, if estimates are to be believed, around four dollars in saved crop yield. These elevated yields, facilitated by pesticide use, help to provide the food on which the global population depends, and at an affordable price.

Pesticides are also widely used to control insects that are vectors of human and livestock diseases. In the 1950s dichlorodiphenyltrichloroethane (DDT) was used extensively against mosquitoes, alleviating malarial risk for millions of people. In Sri Lanka, DDT slashed malarial cases from one million to fewer than thirty by 1964. Yet DDT fell out of favour following the publication of Rachel Carson's *Silent Spring* (1962), a powerful polemic against the profligate use of pesticides, and DDT in particular. Spraying declined, and malaria in Sri Lanka increased once again to around half a million cases by 1969.

Despite the clear benefits of pesticides for food production and disease control, Rachel Carson exposed their insidious environmental and health costs. Famously, she reported how DDT becomes increasingly concentrated in animal tissue up though the food chain. Birds, and birds of prey in particular, suffer from eggshell thinning, failed reproduction, and ultimately population decline. DDT and other pesticides not only damage wildlife, but also endanger human health. Carson's book is credited with launching the global environmental movement.

Pest problems in farming and forestry are at least partly due to intensive production practices where large areas are grown with a single crop or tree species. Monocultures are a bonanza for pests. Increasing crop species diversity is an obvious response to pest problems. It is true that pests also attack plant species in diverse natural communities, but usually the most affected species are very common. The chestnut trees that were wiped out by chestnut blight in the Appalachian Mountains of the United States accounted for around one in four of all trees in these forests. Even tropical rain forest tree species in highly diverse species mixes can occasionally be vulnerable to pests, yet here too it is invariably only the most common species that are affected.

Diversity is therefore not a guarantee against major impacts of pests, but it certainly helps, as many studies have demonstrated.

The question to ask is how species mixes help alleviate pest and pathogen pressures. Ecology has several answers. Mixed species planting implies that individuals of the same species are more widely spaced from each other than if species are grown as monocultures. The progress of a pest or pathogen is slowed if it has to bridge larger distances between susceptible individual hosts. Species mixtures might also challenge pests because the appearance and chemical signature of non-host plants interferes with the pest's host-finding mechanisms.

Natural pest control

Diverse plant communities provide a greater richness of resources and habitat structures that maintain diverse and abundant communities of parasitoids (insects that parasitize other arthropods) and other pest predators. These natural enemies of pests contribute to keeping insect herbivores below economically damaging levels. Natural pest control services provided by native parasitoids and predatory insects in the United States alone provide around $4.5 billion of savings per year in increased yield and reduced insecticide inputs. Integrated pest management strategies aim to maintain healthy populations of such beneficial pest-controlling animals. Farmers are encouraged to use pesticides sparingly, retain flower-rich field margins (Figure 21), hedgerows, and woodland patches in the agricultural landscape, and introduce nest boxes for insectivorous birds.

Whether mixed species planting is a viable strategy for farming and forestry depends on the density of the host in relation to the dispersal behaviour of the pest and the extent to which natural predators are effective in these circumstances. Mixed plantings are economically rather more challenging, due to reduced efficiencies of management and economies of scale. Worldwide, major crops continue to be grown as monocultures, with continued use of chemical pesticides to control ongoing pest

21. Species-rich wildflower strips along agricultural fields, in this case in Switzerland, support a diversity of insects, including many that help to control agricultural pests.

problems, but at substantial costs to beneficial insects such as pollinators and parasitoids. Nonetheless, the concept of integrated pest management, employing a mixture of control techniques, some biological and ecological, others mechanical or chemical, has had wide uptake, especially in tropical agricultural systems. This approach has helped to reduce many pest species to economically tolerable levels, while simultaneously reducing reliance on damaging chemical pesticides.

The enemy of my enemy is my friend

Biological control uses living organisms (or viruses) to suppress specific pest organisms to make them less damaging than they might otherwise be. Alien invasive species might be controlled by importing their natural enemies (parasitoids, predators, or pathogens) from their countries of origin, to establish self-sustaining populations that suppress the invasive pest, or at least limit its spread. Such approaches have the advantage of avoiding repeated and costly applications of pesticides that have lingering environmental impacts.

Biological control has been practised for thousands of years. The first documented description is from southern China, where 2,000 years ago weaver ant nests were encouraged in citrus orchards to control pests. To this day, Chinese farmers create bamboo bridges between citrus trees to encourage the ants to forage throughout orchards. The first deliberate introduction of an exotic natural enemy to control a pest is that of the mynah bird from India, brought to Mauritius in 1762 to control red locusts in sugar cane plantations. Instead of feeding on locusts, the mynahs preferred to eat native lizards that are easier to catch. Indeed, spectacular biological control successes (Box 4) are matched by spectacular failures (as in the cane toad example described in Box 3).

The science of biological control is closely related to that of invasive biology. Charles Elton said as much in his aforementioned book. It concerns pest–enemy interactions and the direct and indirect interactions between populations of target organisms, biological control agents, and resources that humans value. Biological agents effect their control most directly by predating target pest species, but they might also reduce pests by exerting strong competitive pressure on them. Introduced dung beetles successfully suppress bush flies in Australia by being much faster at exploiting and dispersing dung piles, depriving flies of this essential resource. Sterilized red foxes introduced to the Aleutian Islands successfully eradicated the introduced arctic fox through competition, before being themselves removed from the islands.

Viruses have very successfully controlled invasive vertebrates, the most celebrated example of which is myxoma virus used against rabbits in Australia. Evolved resistance by rabbits and reduced virulence by the virus is a classic example of the evolution of intermediate-level virulence that allows disease and host to coexist in reduced numbers. Evolved responses can moderate the long-term outcomes of biological control.

One of the difficulties facing biological control ecologists is that interactions that appear promising in controlled lab conditions often prove ineffective in the field. Unanticipated environmental constraints or unaccounted further interactions affect outcomes. Once released, the success of a novel pest-control species is limited by interactions with other species in the community, or by environmental conditions. Attempts to control water hyacinth (*Eichhornia crassipes*), a globally invasive aquatic weed originally from South America, is a case in point (Figure 22). Introduced

22. **Water hyacinth choking backwaters in Zambia.**

weevil beetles (*Neochetina* species) successfully control water hyacinth on Lake Victoria in East Africa, but not in Florida or South Africa. Much depends on the nutrient levels in the water. High nutrient concentrations increase plant nutritional quality favouring weevil reproduction, but also favour rapid plant growth rates and vegetative spread. Moreover, at low nutrient levels, water hyacinth allocates more resources to flowering than to vegetative growth, thus limiting the food available to weevils. Biological control of water hyacinth by weevils is therefore not effective at very low nutrient concentrations, which can only sustain small weevil populations and encourages plant spread through seed production, nor when nutrient concentrations are high, which is when vegetative growth is prolific. Weevil control of water hyacinth is most effective under modest nutrient concentrations, when weevils have greatest impact in terms of total herbivory relative to compensatory plant growth.

Resilience

Traditional management approaches have sought to regulate the population abundance of organisms, either to maximize resources, be they timber trees, fish harvests, crop yields, or game animals, or to reduce problematic species, including dangerous predators, crop pests, disease vectors, or alien invasives. Such management, referred to as 'command and control', tends to emphasize symptoms rather than causes. Unintended, unexpected, and often catastrophic consequences have resulted from this approach. Such experience has convinced managers of the weakness of command and control, and that resource management is more effective if we consider the whole ecological system. Aldo Leopold recognized this in the first half of the 20th century, but his insights were largely ignored until close to the end of the century. Leopold proposed that to manage populations it is necessary to manage ecosystems, as populations depend on many elements in the ecosystem of which they are a part. Perhaps the complexity of natural systems, and the deficiencies in our knowledge of them, precluded attempts to manage resources as parts of complex systems. This is beginning to change as ecology develops concepts and methods to interpret populations and processes within a more holistic complex systems approach.

The concept of resilience is one element of this thinking, and one that has received remarkable uptake across society. It is also one of those (rather annoying) terms that is given multiple interpretations by ecologists. Generally, resilience is the capacity of an ecosystem to absorb and recover from shocks and disturbances while maintaining overall ecosystem structure and function. More specifically, resilience is the time taken for a system to return to an equilibrium state following a perturbation, or the amount of disturbance that can be absorbed before an ecosystem flips into a new persistent state that is structurally and behaviourally different (Figure 23). All these interpretations assume that ecosystems have relatively stable states to which they tend to return. This is most likely true if we recognize general categories such as broadleaved forest, grassland, or oligotrophic (nutrient poor) lake, while accepting that the species composition of these ecosystems could well change.

A sustainably managed system is one in which its structure, composition, and processes are maintained so as to continue to provide the resources, functions, and services that we value, even when subjected to natural or human perturbations. Delivering sustainability is therefore largely about ecosystem resilience, whereby the array of ecological relationships and functions provide capacity to adapt and recover from perturbations at a range of scales. Ecology theorizes that resilience can be enhanced by maintaining a diversity of species, functional groups, and food web interactions. Managing for resilience also recognizes that disturbance is a natural, even essential, component of ecosystems, which maintains feedbacks and flows across scales, and contributes to the adaptive capacity of ecosystems. Understanding

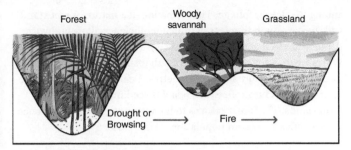

Forest   Woody savannah   Grassland

Drought or Browsing ⟶   Fire ⟶

23. A representation of alternative stable states. Basins represent alternative stable ecosystem states, each with their own resilience, reflected by basin depth. The current state might be imagined as a ball lying at the bottom of a basin. Perturbations shift the ball in one direction or another up the slope. If the perturbation is not great, the ball settles back to the bottom of the basin, reflecting the recovery of the ecosystem to its stable state. Larger perturbations that exceed the ecosystem's resilience, represented by the height of the peaks, cause our imaginary ball to fall into a neighbouring basin, implying a shift to an alternative stable ecosystem state.

this requires appreciating the dynamic nature of ecosystems in time and space, reflected in cycles of change and patchiness.

Dynamic stability

The dynamic nature of ecosystems must be reconciled with their purported stability. Changes within ecosystems, characterized by population fluctuations, shifts in food web interactions, or variations in resources flows, allow ecosystems to adjust to disturbances. Such changes might appear dramatic, but are often stages of natural adaptive cycles. North American spruce-fir forests, widespread and seemingly stable ecosystems, are the habitat of the eastern spruce budworm (*Choristoneura fumiferana*), a moth whose caterpillars feed on spruce and other coniferous trees. Birds feed on the budworm keeping its populations low when forest stands are young and the canopy relatively open. As trees mature, birds find it increasingly difficult to find budworms among the denser foliage. The formation of a mature forest stand and the corresponding loss of predator efficiency allow budworm populations to increase rapidly, causing a budworm outbreak. Widespread defoliation and tree death results, with mortality extending over many square kilometres. Dead trees slowly release their nutrients into soil, which benefits seedlings that colonize dead stands. The forest returns to the young stage, and birds once more control budworm populations (Figure 24).

The forest–budworm cycles are dynamic, yet stable in that the system remains a spruce-fir forest. There are long-lasting local equilibria, when birds keep budworm populations low, and thresholds, the point at which birds are no longer able to control budworm populations. Dramatic collapses, expressed by widespread tree mortality, are followed by renewal through seedling establishment and succession. At all times the forest is still recognizable as a forest, albeit in different stages of an adaptive cycle.

Ecosystems are patchy. The cycle of maturation, persistence, catastrophic change, and regeneration is not synchronized across their entire extent. Different areas of an ecosystem occupy different stages of the adaptive cycle at any one time, giving rise to a landscape mosaic of patches. The patchiness created by a complex disturbance regime provides resilience. The flora and fauna of mature phases persist in undisturbed patches, while those associated with regenerating phases relocate from one short-lived regenerating patch to another elsewhere in the landscape. Remnant mature patches act as seed sources to initiate the pioneering phase of recovery. Patchiness also limits the impact and extent of disturbances, which largely affect mature phases in the cycle.

Disturbances themselves are patchy in their severity and extent. Even very large occasional disturbances have heterogeneous landscape impacts. Some patches might escape a disturbance by

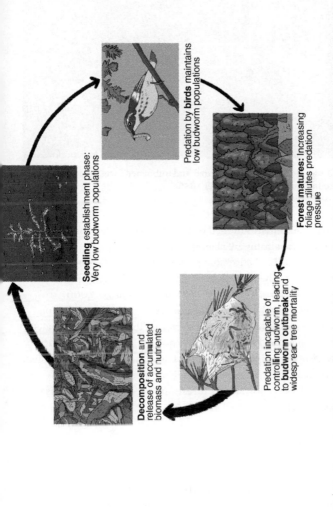

Seedling establishment phase: Very low budworm populations

Predation by **birds** maintains low budworm populations

Forest matures: Increasing foliage dilutes predation pressure

Predation incapable of controlling budworm, leading to **budworm outbreak** and widespread tree mortality

Decomposition and release of accumulated biomass and nutrients

24. The dynamics of the spruce-fir forests of North America. The arrows indicate the sequence of steps in the dynamic cycle between different ecosystem stages. The width of the arrows represents the rapidity of change, wider arrows representing more rapid change.

113

25. A mosaic landscape of burned and unburned forest patches after the Yellowstone fires in October 1988.

virtue of local environmental conditions that render them less vulnerable, or simply by chance. The summer of 1988 was the driest on record in Yellowstone National Park. It did not take much to ignite the lodgepole pine forests, and the extent and severity of the fires were unprecedented. Around 570,000 hectares of forest burned. In the midst of a burn, the landscape must have seemed desolate. Yet surveying the aftermath revealed a mosaic of burned and unburned patches (Figure 25). Rapid recovery of the lodgepole pine forest ensued, seedlings established, and animal populations recovered. Now, some three decades after the fires, Yellowstone's forests are regenerating as dense stands in formerly burned areas.

Alternative stable states

The forests of Yellowstone have coexisted with fire for millennia, and have high natural resilience to fires, even those as severe as the 1988 event. Large fires, recurring at intervals of 100 to 200

114

years, burn through the forest canopy and kill mature trees, but they release nutrients and allow light to reach the forest floor. This triggers seed germination and seedling growth. The forest recovers. Yet it is not obvious that such resilience will be maintained into the future. Climate change might push the ecosystem beyond its historic ecological experience and drive it towards an alternative stable state. The exceptional hot and dry weather of 1988 in Yellowstone is no longer exceptional. Forests that are adapted to occasional large severe fires are now prone to frequently recurring fires. These could burn forests before they are able to recover. Ecologists studying fire regimes think it plausible that the nature of interactions between fire, climate, and vegetation expected by the middle of the 21st century will no longer be amenable for the persistence of Yellowstone's coniferous forests. New fire regimes could transform the flora, fauna, and ecosystem processes in this and similar forests in North America, leading to the replacement of the conifer forests with a non-forest plant community.

Documented examples of transformational ecosystem shifts (including that of the collapse of the North Atlantic cod fishery) lend credibility to projections of Yellowstone's future. African savannahs support an assemblage of fire-maintained grasses that limit the extent of woody vegetation. Transition to a more wooded alternative state occurs if grazing pressure is sufficient to reduce grasses substantially. This increases establishment by woody plants, and reduces fire frequency as grass fuel loads decline, ensuring the woody community persists for decades. Large browsers, such as elephants, can, eventually, re-establish grassy savannahs by creating openings in the canopy and physically breaking trees (see also Figure 23).

Socio-ecological systems

Failure to predict alternative states, or anticipate transformational ecosystem change, can be very costly, as the collapse of fisheries illustrates. It is imperative that ecosystem management moves away from a command and control approach to the management of single resources to one that considers species interactions, resilience, ecosystem dynamics, and adaptive cycles. Ecosystem managers consider how, and to what extent, human-induced changes affect ecosystem resilience, and whether resilience can maintain stability within acceptable boundaries of change. Strategies to secure resilience can be diverse. A water company might maintain drinking quality water in a lake by reintroducing a top predator, removing non-native fish that disturb sediments, lobbying for policies to reduce fertilizer run-off into watercourses, or encouraging landowners to maintain riverbank trees to minimize erosion and support healthy aquatic invertebrate communities. Collectively, these actions enhance the resilience of the lake by maintaining diverse food webs and reducing biophysical changes that might cause the lake ecosystem to flip to an undesirable alternative stable state. Such actions require collaboration across different stakeholders, from the water company to upstream landowners, policy-makers, and other interest groups such as fishermen and conservation bodies.

These challenges require ecosystem managers to be proficient in applied ecology, and also familiar with social and policy issues. Human and natural realms in ecosystem management are closely intertwined. Applied ecologists contribute important knowledge on how ecosystems function, but ecosystem management takes this a step further by incorporating this knowledge within a larger socio-ecological framework of reciprocal feedbacks between natural and human systems. Applied ecologists need to be comfortable working alongside social scientists, economists, behavioural psychologists, policy-makers, and landowners, among many others.

Chapter 7
Ecology in culture

In the public conscience, the science of ecology is often conflated, much to the annoyance of academics, with an eclectic spectrum of naturalists, poets, organic farmers, birdwatchers, and activists, all of whom care deeply about nature and the environment. 'Ecology', as a concept and a term, has become adopted and adapted by a variety of cultural contexts and purposes. By being so, it has become politicized and value laden. Social ecological perspectives that underlie the growth of the modern environmental conscience are, and have been, inspired by ideas emerging from the science of ecology, notably those of interdependency, holistic thinking, resilience, and adaptive systems. Moreover, some primarily ecological disciplines, conservation biology pre-eminent among them, are inherently value laden and, as such, have been influenced by the development of modern cultural values concerning the relationship that human society has (or should have) with the environment. It is therefore not always easy to differentiate the science of ecology from its broader social interpretations, particularly when framed by current environmental challenges.

The cultures of ecological thought draw widely from historical precedents, often gaining inspiration from aboriginal ethics and practices, whether real or imagined. The speech of Chief Seattle of the Divamish Indians is a case in point, in which he purportedly declaimed the white man's estrangement from nature: 'This we know: the earth does not belong to man, man belongs to the earth. This we know. All things are connected, like the blood which unites one family.... Man did not weave the web of life, he is merely a strand in it. Whatever he does to the web, he does to himself.' The contested historical and literary accuracy of Chief Seattle's speech matters little. More to the point is that ecological ideas of interdependency and holism preceded the science of ecology itself. Ecological science has formalized an understanding, the essence of which has been known for centuries by cultures around the world. Yet ecological science has also informed and inspired a modern ecological conscience, one that responds to new environmental challenges that are qualitatively and quantitatively different from anything that humans have experienced previously. Indeed, perhaps the most important cultural revolution of the 20th century was the transfer of ecological insights of holism, feedbacks, and inter-dependencies, from scientific to moral and political fields.

Ecological conscience

Ecology is more than a science, it has become a worldview. In *The Ecological Conscience*, a speech delivered on 27 June 1947 to the Conservation Committee of the Garden Club of America, Aldo Leopold emphasized, 'Ecology is the science of communities, and the ecological conscience is therefore the ethics of community life.' This presaged Leopold's 'Land Ethic', which encompasses connections between ecology, ethics, policy, and management action. There has always been tension between ecologists who readily advocate conservation, and those who see advocacy as inappropriate in science professions. Aldo Leopold was firmly on the side of advocacy. Others have similarly despaired of an overly reductionist and narrow disciplinary ecological approach. In *The Conduct of Life* (1951), Lewis Mumford wrote, 'So habitually have our minds been committed to the specialized, the fragmentary, the particular, and so uncommon is the habit of viewing life as a dynamic inter-related system, that we cannot on our own premises recognize when civilization as a whole is in danger.' This mid-century awakening of ecology to environmental concerns was given added impetus by Fairfield Osborn's *Our Plundered Planet* and William Vogt's *Road to Survival*, both published in 1948. The publication of Rachel Carson's *Silent Spring* (1962) is credited as the watershed moment when ecology became a politically and culturally embedded science.

Paul Sears, writing only two years after the publication of *Silent Spring*, called ecology 'the subversive science'. Sears was quite clear that the principles of ecology threatened many assumptions and practices of the existing socio-political order. Ecology appeals to movements seeking to realign social, political, and economic systems to agendas that give centre stage not only to environmental concerns, but also to those of social injustice and inequality. The Marxist thinker Murray Bookchin argues that the exploitation of nature arises from unjust social frameworks, and that just social frameworks are ecologically sound. Ecological feminism argues that patriarchy and anthropocentrism, the human domination of nature, are two expressions of the logic of domination. An environmental ethic that dismantles anthropocentrism will, by similar logic, weaken patriarchy. The Deep Ecology movement follows similar thinking in ascribing environmental ills that afflict modern society to the sundering of the individual egoistic self from the larger biological whole. Deep Ecology appeals to the reconstitution of the inextricable relationship between individual and biosphere.

These interpretations of ecology are far removed from the science of ecology, and yet draw from the same pool of ideas that ecological science has generated. Ecological concepts, interpretations, and philosophies have been adapted for diverse purposes in many social and political discourses. Most notably, ecology provides the theoretical and conceptual justification for the normative stance of biological conservation. It also lies at the heart of Green politics, which emphasizes the co-dependency of humans with their environment. This echoes John Muir's oft misquoted remark: 'When we try to pick out anything by itself, we find it hitched to everything else in the Universe' (*My First Summer in the Sierra*, 1911). More prosaically, ecology has been adulterated to serve the needs of marketing and advertisement, purportedly reflecting the environmentally benign qualities of 'ecological' products. In 1971, for example, Coca-Cola Company advertising referred to its recyclable glass bottle as 'the bottle for the Age of Ecology'.

Gaia

Alongside mainstream ecological science, emphasizing individual components and their relations, is another view of the Earth as an emergent product of all its species and their interactions, with a coherence that justifies its description as a living organism. Gaia Theory was developed by James Lovelock, an atmospheric chemist, and Lynn Margulis, an evolutionary biologist. Simply stated, they argue that the Earth is alive in the sense that it is a self-organizing adaptable entity derived from the interactions of organisms with each other and their geological, marine, and atmospheric environments. This living organism, Gaia, self-regulates essential aspects of its environment, such as its temperature, ocean salinity, and concentrations of atmospheric oxygen and carbon dioxide.

Lovelock has been at pains to emphasize that Gaia is not teleological, in that its self-regulation processes have no purpose to benefit life, but life benefits nonetheless. The Earth's temperature, for example, depends on the concentration of atmospheric carbon dioxide. Volcanic activity is the only substantial natural source of atmospheric carbon dioxide, and the main sink is rock weathering, which combines calcium with carbon dioxide to form calcium carbonate that is eventually deposited onto the sea floor. Weathering is greatly accelerated by bacteria and plants that actively transfer carbon dioxide from atmosphere to soil. In the oceans, marine algae and corals accelerate sea floor deposition of carbonates by sequestering dissolved carbonates into chalky shells and coralline reefs. Accumulated carbonate sediments lock away carbon dioxide as deposits of chalk and limestone. Some marine algae, the coccolithophores, also contribute to the formation of clouds. When they die, they emit gaseous dimethyl sulphide, which emerges into the atmosphere above the ocean surface to produce tiny acid droplets. Water vapour condenses on these droplets to form clouds, which in turn reflect the sun's energy. Thus life is an essential component of the global feedback systems that contribute to the regulation of atmospheric carbon dioxide and, consequently, global temperature.

Gaia theory has inspired thinking that goes well beyond mechanistic processes and outcomes, leading to the birth of the new scientific discipline of Earth Systems Science. The image of the Earth from space, first photographed in its entirety in 1972, had a forceful impact in underscoring a holistic (and literal) worldview (Figure 26). Regardless of Lovelock's original intentions, Gaia in popular culture has thus come to represent a holistic worldview of ecology and our interactions with Nature that stands apart from the reductionist scientific approach that characterizes much of ecological science. Gaian thinking has gone beyond Lovelock's original intent by incorporating metaphysical or spiritual interpretations. Recognizing that Gaia means different things to different people, proponents are attracted to the Gaia concept by the idea that we, as humans and individuals, are an integral part of a larger entity from which arises a deep appreciation of the reality of interdependence. This justifies and motivates opposition to environmental degradation, and it is likely for this reason that Gaia has captured the imagination of many individuals and environmental and conservation movements.

26. The Blue Marble. This image, first photographed in 1972, did much to communicate a holistic view of Earth as an entire, and finite, planetary community.

Deep Ecology

Closely related to Gaia, at least as popularly interpreted, is the sense that the entirety of what has arisen through millions of years of evolution has value and meaning that is impossible to articulate through reductionist scientific research. This sensitivity characterizes the Deep Ecology movement, based on the writings of the Norwegian philosopher Arne Naess, who coined the term. Deep Ecology asserts that the biosphere does not consist of discrete entities, but of internally connected and interacting individuals that together constitute a fundamental reality.

Adherents of Deep Ecology castigate the human-centred individualism at the core of Western culture. Deep ecologists argue that environmental philosophy must recognize the inherent values of nature, independent of human needs. They reject mainstream environmentalism predicated on human welfare priorities, rather than intrinsic values. By accepting the equal and intrinsic value of all organisms, and recognizing our ecological unity, deep ecologists claim a deeper understanding and connection to the natural world.

Deep Ecology is, in principle, an egalitarian philosophy, with equal moral weight being assigned to all biota. It has been widely critiqued because egalitarianism leaves little scope for ethical decision-making. If all organisms have equal worth, then how do we adjudicate among competing interests? On this basis, in 1980, Baird Callicott argued that environmental ethics cannot 'accord equal moral worth to each and every member of the biotic community'.

Cultural ecology

Applying ecological thinking to human societies provides insights on our norms, rituals, and taboos. Cultural ecology argues that the natural environment contributes substantially to societal culture and organization. The anthropologist Julian Steward envisioned cultural ecology as the 'ways in which culture change is induced by adaptation to the environment'. Steward argued that the environment influences human culture, rather than determining it, though later proponents of cultural ecology have been critiqued for proposing stronger environmental determinism of culture. Steward's fieldwork among the Shoshone people in North America emphasized how complex cultural strategies enabled them to live in the desert terrain of the Great Basin between the Sierra Nevada and Rocky Mountain ranges. Their detailed knowledge of seasonal variations in the availability of resources as diverse as pine nuts, grasses, berries, deer, elk, sheep, and antelope shaped Shoshone culture by influencing their patterns of migration, their social interactions, and cultural belief systems.

One of Steward's most eminent students, Roy Rappaport, went on to study subsistence practices of the Tsembaga in Highland New Guinea. Rappaport used ecological concepts such as energy flows, carrying capacity, and mutualism to explain the rationale behind the management of resources by the Tsembaga. Pigs in Tsembaga villages cleaned up the village wastes and grubbed up weeds from fruit tree orchards, but they began to create problems when their populations became too high. The Tsembaga people used periodic ritual feasts to reduce the pig population to ecologically appropriate levels. In using ecological concepts to understand the Tsembaga subsistence practices, Rappaport downplayed the role of cultural beliefs in favour of ecological constraints.

Similarly, Marvin Harris applied functional and materialist thinking to Hindu beliefs about sacred cattle in India. Among Hindus, cows are venerated as symbolizing divine and natural beneficence, and consequently eating beef is avoided. Harris argued that such cultural belief is perfectly rational within Hindu ecological and economic systems. Restrictions regarding cows, he argued, respond to the need for milk, dung for fuel and fertilizer, and labour for ploughing.

Studying cultural practices through ecological filters provides a more sympathetic and informed understanding of environmental management. For many years, shifting cultivation was castigated by scientists, environmental activists, and the media, for driving tropical deforestation. Small-scale farmers who practise shifting cultivation clear and burn a small patch of forest in which they plant annual and perennial crops such as rice, beans, corn, taro, and manioc. Over time, soil fertility declines and insect pests accumulate. After a few years, farmers introduce fruit trees into the plot, and clear new forest areas for their crops. The cycle of forest clearance, cultivation, and abandonment was interpreted as destructive and wasteful by environmentalists, who succeeded in convincing governments in some tropical countries to prohibit shifting cultivation practices in favour of settled agriculture. Cultural ecology applied a different interpretation to shifting cultivation, by recognizing the detailed environmental knowledge possessed by cultivators that has sustained their practices over many generations. Shifting cultivation is, at relatively low human population densities, highly sustainable, as cultivation is followed by long fallow periods that rebuild soil nutrients. Planted fruit trees encourage birds and rodents that bring in seed of other tree species from surrounding forests, and further enhances forest recovery. The clearance of relatively small patches mimics natural processes of tropical forest disturbance, in which storms and tree falls periodically open up small areas. Small cleared patches even enhance biodiversity by creating a greater variety of habitats in a given area.

Inappropriate environmental perspectives derived from colonial legacies of land management that took no account of traditional management practices have led to erroneous policies. James Fairhead and Melissa Leach's work in West African Guinea describes how state forestry officials, influenced by Western notions of land management and ecological transition, interpreted mosaics of forest patches as remnants of more extensive tropical forests. The government officials assumed that local people had destroyed the forest, and so imposed regulations and fines to prevent further forest loss. Fairhead and Leach showed that forest islands were, in fact, expanding due to tree planting and fire control activities undertaken by local people.

The noble savage

Overly naive cultural ecology feeds into the image of the ecologically noble savage, a historically persistent myth of Western culture. The modern myth of the noble savage is commonly attributed to the Enlightenment philosopher Jean Jacques Rousseau, though he never actually used the phrase himself. Romantic writers and artists of the 18th century and since have been captivated by the idea of an idealized, simpler past, in which people lived harmoniously with nature. Indigenous societies were celebrated in art and literature, and in scholastic works, as being culturally and spiritually attuned to the ecology of their environment. The continuing purchase of these ideals is evident in James Cameron's 2009 film *Avatar*, in which the native Na'vi sustain a vibrant natural world.

These ideas have modern counterparts. The Kayapó in the Amazon, and Eastern Penan in Bornean rain forests, have both stood in opposition to the clearance of forests, building of roads, and construction of dams, in defence of their own traditional territories and forest-based lifestyles (Figure 27). These indigenous communities have organized, sometimes aided by environmental organizations, to protect their forests, and have successfully halted some large-scale development projects.

27. **Penan protesting against incursions by loggers into the forests on which they rely.**

The combination of rain forest conservation and indigenous rights has been immensely effective in securing media attention, appealing to the longstanding Western idea of an environmentally noble savage, while aligning to growing environmental concerns.

Ideas that attribute an ecological sensitivity to indigenous people have been critiqued as misrepresenting traditional societies, and even undermining their legitimacy. To suggest that indigenous people have had no impact on their environment is to deny their human history. Many indigenous societies have, indeed, greatly and permanently changed their environment. Indigenous people have actively managed and modified landscapes for thousands of years. The aboriginal Australians have done so for 60,000 years, using fire to modify landscapes that quite possibly contributed to the extinction of many of Australia's megafauna. Fire likewise has been used by Native Americans, and likely also contributed to species extinctions.

Moreover, indigenous communities often do acquiesce and even support development programmes that run counter to the cultural expectations projected onto them. Much to the horror of conservationists, the Kuku Yalanji people supported the construction of a road to the south of Cooktown in Cape Tribulation, Australia. Sarawak's Western Penan, in contrast to their Eastern Penan neighbours, actively seek compensation agreements with logging companies for access to their lands, and welcome the associated benefits and employment. They do not welcome logging, but rather, like any other community, they adapt to make the best of the existing conditions.

Communities, regardless of their traditions and cultures, will tend to use resources sustainably when population densities are low, and access to markets and technology is limited, rather than due to any particular conservation ethic. This is not to say that conservation ethics are not important, but rather that they are subsumed by greater priorities of securing individual wellbeing.

As human populations grow, they will eventually exceed the carrying capacity that can be supported by local environments. This creates ecological crises that can lead to warfare, migration, or social collapse, or to institutional and cultural change towards new social and production systems and philosophies.

Sacred ecology

While the noble savage idea is no longer tenable in its original form, there is much interest in understanding how traditional societies, with a long history of local land and resource use, interpret the environment around them. Traditional ecological knowledge is the body of knowledge, practice, and belief, accumulated and adapted through generations by cultural transmission, about the relationships of humans with other organisms and their environment. Mostly, societies that have a rich traditional ecological knowledge have a more direct relationship with their environment, and are less technologically oriented. Many are indigenous or tribal, but not necessarily so.

Fikret Berkes recognizes a 'sacred ecology', a strong belief component within traditional ecological knowledge, which shapes peoples' perceptions of how they interact with nature and its elements. This moral context makes it impossible for these societies to differentiate religion from ecology. Ecological attributes cannot be separated from social or spiritual aspects. Stories and rituals signify and communicate ecological meaning that imbues a 'sense of place'. For the Penan of Sarawak in Malaysian Borneo, Peter Brosius explains that 'The landscape is more than simply a reservoir of detailed ecological knowledge…It is also a repository for the memory of past events, and thus a vast mnemonic representation of social relationships and of society.'

The disassociation of culture and ecology in industrialized societies has led to, or perhaps is the result of, the replacement of traditional environmental management with industrialized and production-oriented management systems. It is right to critique the myth of the noble savage, but it is equally important to acknowledge the validity of other forms of ecological knowledge that have long stood the test of time. To restore environmental health through renewed environmental awareness, we perhaps need to rebuild a cultural ecology to inspire an urban industrialized global population.

Chapter 8
Future ecology

Being an ecologist used to be simple. The ecologists' toolkit of little more than a pair of binoculars, a tape measure, perhaps some sampling traps and tubes, and a notebook, has proved effective in probing the most complex natural systems. Yet ecology in the 21st century has entered the era of Big Data and new technologies. Ecologists routinely use satellites, drones, tracking devices, genetics, and stable isotopes to monitor and interpret patterns, processes, and interactions across multiple spatial and temporal scales.

Ecological questions used to revolve around issues that had little direct bearing on politically charged concerns. This also has changed. Ecologists now routinely deal with issues of conservation, land management, and resource use, topics that are inherently normative and often conflictual. We reside in the Anthropocene epoch, defined by humanity's permanent legacy on the Earth system. This legacy includes biodiversity losses, climate change, elevated atmospheric carbon, ocean acidification, nitrogen deposition, the spread of invasive species, deforestation, and soil erosion, all of which alter ecosystem structure and functioning for the long term. In these circumstances, ecological understanding built on natural and largely undisturbed systems of the past provides limited guidance for future patterns and outcomes.

There are many uncertainties about how populations, communities, and ecosystems will respond to global change. Elevated atmospheric CO_2 might increase plant production through a fertilization effect, but could equally render plants vulnerable to drought or nutrient deficiencies. Ocean acidification will affect the productivity of marine ecosystems and the structural integrity of coral reefs, with potentially disastrous implications for marine biodiversity and fisheries. Global warming is facilitating the spread of pathogens and pests into new regions, causing destructive pest outbreaks in American forests. The cascading consequences of these outcomes through food chains and communities, and across ecosystems, is uncertain. Neither do we properly understand the factors determining ecosystem resilience, or even how to measure resilience. In view of ongoing and rapid losses of biodiversity globally, we are worryingly ignorant of how genetic and species diversity, and the structure of ecological interaction networks, affect ecosystem functioning and resilience. There is much to learn and do.

Uncharted territory

Global environmental changes affect the frequency and severity of disturbances such as wind, droughts, fire, and pest outbreaks. The coming decades will experience novel disturbance regimes, and more frequent extreme events. The ability of ecosystems to respond to these changes is influenced by their resilience, itself affected by human activities that alter ecosystem composition and structure. A comprehensive understanding of trajectories of environmental change, and ecosystem responses, has become a central purpose of ecology. Of particular concern is whether ecosystems will cross thresholds or tipping points, leading to transformative shifts from which recovery is difficult or even impossible on human time scales.

As we move into this uncharted territory, the past is no longer an effective guide to the future. Nonetheless, ecological

28. Forest recovery after eruption of Mt St Helens in 1980.

understanding of how ecosystems function is invaluable in anticipating the processes and outcomes of environmental change, and planning the appropriate responses to it. Ecosystems can recover, often quickly, from even large natural disturbances. The recovering forests on the slopes of Mt St Helens following the volcanic eruption in 1980 are a case in point (Figure 28). Yet we should not be complacent. Ecosystems will have to deal with future conditions and disturbance regimes that they have never previously experienced and to which they are not obviously adapted (Box 5).

The cumulative and synergistic effects of recurring disturbances also threaten habitats. Amazonian rain forests do not burn naturally, yet incursions of roads and farms, coupled with climatic changes, are beginning to dry the forest interior, and the presence of humans increases fire frequency. Initially, fires have low intensity, but rain forest trees are not adapted to fire, and even low-intensity fires kill many of them. Dead trees create openings

Box 5 Beetles, blisters, bears, and the story of the whitebark pine

Forests of whitebark pine (*Pinus albicaulis*) in the northern Rocky Mountains are under attack by white pine blister rust, a non-native fungal pathogen (*Cronartium ribicola*), and the native mountain pine beetle (*Dendroctonus ponderosae*). Beetle distributions were formerly limited by cold temperatures at the high elevations occupied by whitebark pine, while the blister rust, originally from Asia, was introduced into North America around 1900. Whitebark pine has not evolved defences against the beetle or the blister rust as it has never been exposed to these threats until now. Other conifers will most likely replace lost whitebark pine in due time, but there are likely to be cascading effects on grizzly bears and other animals for which whitebark pine seeds are an important food source. Moreover, extensive whitebark pine death can increase wildfires by creating extensive stands of dead trees.

in the canopy, which further dries understorey layers, and accumulating dead wood raises fuel loads. This sets the stage for increasingly severe and extensive fires, especially when coinciding with more frequent drought events. Some ecologists believe that these synergistic disturbance interactions threaten to transform moist rain forest into much drier wooded savannah. Should such scenarios unfold, the consequences for biodiversity and carbon emissions are immense.

Tracking climate

Climate change represents perhaps the most substantial long-term threat to biodiversity and ecological communities. One study has estimated that 15–37 per cent of all species are vulnerable to climate change-related extinction. As the climate warms, species that are not able to adapt quickly have little option other than to move to regions more suited to their climatic needs. Modelling studies estimate species' future expected ranges under different climatic scenarios, but it is not yet clear if species will be able to disperse sufficiently quickly given rates of climate change.

Plant, mammal, bird, and butterfly populations have already shifted to higher latitudes and elevations in response to climate change. Many species have responded rapidly, but others have not been able to keep pace with geographical temperature shifts. From a sample of 35 non-migratory European butterflies, 22 species experienced northerly range shifts of 35–240 km during the 20th century, over which period climatic isotherms have shifted northwards by around 120 km. The remaining 13 species have moved little. Many birds and butterflies, and even those with high individual dispersal capabilities, have accumulated substantial 'climate debt', the lag between species' actual colonization of new areas and the rate required given the pace of climate change. One study suggests that between 1990 and 2008, the northward temperature shift in Europe has been sufficiently rapid to incur average climate debts of 212 km for birds and 135 km for butterflies. If such studies are correct, then even highly mobile bird and butterfly groups might fail to keep up with climate change.

Bumblebees appear particularly vulnerable. Most species have failed to disperse much at all beyond their current northern range limits. They have meanwhile suffered contractions of their southern range limits by as much as 300 km in both Europe and North America, most likely due to more frequent anomalously high temperatures. Dispersal of reproductive queen bumblebees typically ranges from three to five kilometres per year, but infrequent longer-distance dispersal events also occur, so dispersal capabilities do not appear to be limiting. Instead, it might be that incoming species face competition for space and resources with already established species. Bumblebees expanding their range northwards might also encounter plant communities that are less rich in nectar and pollen providing flowers. Tracking climate change is not simply a matter of species range shifts, but also of wholescale community readjustment and organization. This is a ripe area for future ecological study.

Many species at risk of climate-related extinction are those with narrow distributions and limited dispersal capabilities. One option to maintain wild populations of these especially vulnerable species is to translocate them to new suitable locations. Bumblebees can be easily translocated by relocating small numbers of fertilized queens in the spring to climatically suitable habitats. This kind of assisted colonization has been criticized for being overly interventionist, leading to the creation of 'unnatural' communities. The counter-argument is that conserving biological communities as they are now or have been historically is not tenable given climatic and other human-driven environmental changes. Translocation has precedents in interventions to safeguard threatened species, or to replace extinct species with ecological analogues. Critically endangered New Zealand birds, the flightless kakapo parrot and the takahe, a flightless rail, were introduced to predator-free island refuges outside their native range to aid their recovery. Giant Aldabra tortoises were introduced to Round Island, Mauritius, in the Indian Ocean, to restore seed dispersal of the native ebony trees, a function formerly performed by extinct native tortoises. Advocates of assisting the movement of organisms into climatically suitable new ranges argue that such actions are simply incorporating climate change into existing current conservation frameworks.

Pleistocene parks and Frankenstein ecosystems

Restoring lost biota by reintroducing large mammals and birds as herbivores and predators into habitats where they were once present, be they wolves in Yellowstone or beavers and white-tailed eagles in Scotland, known as 'rewilding', has become a popular rallying cry for conservationists. It is also controversial. Detractors

29. A representation of a rewilded landscape.

raise concerns about the impact and suitability of reintroducing animals into regions from which they have long been absent. Landscape rewilding takes this a step further, by arguing that extensive areas should be restored to allow the recovery of former ecosystems, self-sustained by a biodiversity similar to what existed prior to major human intervention (Figure 29). Such ideas have been termed Pleistocene parks or, more disparagingly, Frankenstein ecosystems. Rewilding aims to recover ecological interactions that were lost on the disappearance of large fauna that existed before wholesale transformation of land and habitats by humans. Rewilding advocates aim to rebuild ecosystems by removing infrastructure, avoiding active management of wildlife populations, promoting natural forest regeneration, and reintroducing lost native species. Some even argue for the introduction of large African and Asian mammals to the American continent. Sceptical ecologists retort by questioning the appropriateness of reintroducing species into landscapes that have undergone substantial transformation and readjusted to new realities. Moreover, the larger problem, they contend, is the risk of new and unwanted ecological interactions arising.

Local people have legitimate concerns about sharing a landscape with large wild animals with which they have little familiarity. Wild boar were once common in the UK, but were exterminated in medieval times. Imported from continental Europe in the 1980s to farm for meat, they promptly escaped and established wild breeding populations by the early 1990s, much to the consternation of farmers whose crops they damage, and recreational walkers who feel intimidated by free ranging wild pigs. Setting aside questions about what people want and are prepared to accept, rewilding needs to respond to the ecological issue of whether current landscapes continue to retain the niche space necessary to support reintroduced species. Conservationists need to undertake ecological due diligence to ensure that reintroductions have a chance of succeeding without causing inadvertent damage through unforeseen ecological effects.

Beyond evaluating species' ecological requirements, ecologists must consider the possible changes to a habitat resulting from successful reintroductions, including implications for other species. Organisms modify their environment, and in doing so they affect the conditions and resources available to other organisms sharing the same habitat. Changes can be either positive or negative, and often both. When beavers build dams they modify nutrient cycling and decomposition processes, change the physical structure of river systems, influence the quantity and type of materials transported downstream, and shape plant composition in riparian zones (Figure 30). Their reintroduction in Scotland in 2009 has mostly been beneficial for local biodiversity, and for mitigating floods and droughts, but each species slated for reintroduction needs to be treated, and studied, on its own merits.

30. A dam built by Eurasian beavers (*Castor fiber*) creating a shallow pond in marshland, Tayside, Scotland.

Ecological restoration

Approximately 25 per cent of global land area has been degraded through soil erosion, salinization, peatland and wetland drainage, forest loss, or desertification. This is not a new problem. Histories of degradation and landscape transformation extend back thousands of years, as evidenced by archaeological remains and charcoal in soil layers in even remote areas of temperate and tropical zones. Over two millennia ago, Confucius recognized and described soil and vegetation degradation in the East, as did Plato and Aristotle in the West. Jared Diamond argues, albeit controversially, that environmental degradation has caused the decline and collapse of several human civilizations in history. In the mid-20th century, Aldo Leopold and Rachel Carson, to name but two of many, heralded a new era of environmental responsibility by recognizing the need to conserve and restore landscapes and habitats that we have degraded. The idea of restoring landscapes has received widespread impetus from a broad global movement to plant trees to sequester carbon for climate change mitigation, and to recover the ecological function and biodiversity of degraded forest systems. The Bonn Challenge has set a target to restore 350 million hectares of degraded land back to forest by 2030. Tree planting is deeply rooted in many cultural norms, and public enthusiasm for restoration through tree planting is unsurprising.

Yet restoring landscapes is not only about planting trees. Forest and landscape restoration interprets landscapes and the ecosystems they contain as complex adaptive systems, consisting of many components that interact across multiple spatial and temporal scales. This is a holistic ecological perspective, albeit one that is not without challenges for future restoration planning. Interactions across scales can dampen or amplify environmental fluctuations, giving rise to dynamic and often unpredictable outcomes. In young regenerating fallows in the central Amazon, prior management intensities influence recovering forest structure at local scales, whereas the make-up of the surrounding landscape, a much larger-scale effect, determines the diversity of forest species. There are many possible outcomes of regeneration processes, subject to local and landscape scale interactions.

Given multiple trajectories of recovery, and uncertainties about future climatic and environmental conditions, a more appropriate goal of landscape restoration might be to bolster ecosystem resilience, rather than restore particular ecosystem structures or compositions. Building ecosystem and landscape resilience depends on (re-)establishing functional processes, including plant–soil interactions underlying carbon and nutrient cycling, predator control of trophic networks, and pollination and seed dispersal. This requires coordinated action among landowners, landscape managers, planners, and policy-makers. Policy processes now take heed of 'natural capital', the environmental assets that directly or indirectly have value for people, and include species, freshwater, forests, soil, air, and oceans, as well as the ecological processes and functions that link these components and sustain life. The future management and restoration of landscapes will undoubtedly place new demands on ecological expertise.

New technologies

Ecology has become broader in its spatial outlook. Environmental assessment and ecological monitoring, for landscape restoration or other purpose, requires observation of population dynamics, species interactions, ecosystem states and flows, and disturbance impacts, across multiple spatial scales. Ecologists are benefiting from new technologies for this purpose, and the advent of Global Positioning Systems (GPS), satellite communications, remote sensing, and high speed computing, and the genetic revolution have transformed ecological science.

Ecologists now track animals more extensively and accurately than ever before. Tracking devices include accelerometers, similar to those in fitness monitors used for sports, which provide information on animal movements, behaviours such as sleep or prey capture, and metabolism such as heart rate and energy use. Miniaturization permits the use of trackers on fish, birds, and even insects.

Remotely activated cameras triggered by motion sensors, known as camera traps, have long been used to record wildlife, but their utility is limited by the need to visit each camera to manually download data. By linking camera traps into a wireless sensor network (WSN) pictures can now be downloaded remotely. Any other type of ground-based sensor configured to a WSN can collect data locally, and transmit the data via the WSN to a central data hub, from which information can be uploaded. Data from tiny micro-climatic sensors, deployed across landscapes by the thousands, can be efficiently shared using local wireless systems so that only one device need be collected to access the information obtained by many. These technologies reduce the need for multiple and arduous trips to each sensor location for manual downloads, and have the added benefit of minimizing disturbance to animals or sensitive areas.

Listen!

We are a visual species. Take a moment to close your eyes and listen to the world around you to discover a different perspective on nature. Ecologists are beginning to record the sounds of landscapes, or soundscapes, from which they extract useful information about species composition and ecosystem complexity. The loss of environmental acoustic richness reflects human environmental impact, as portrayed by the film *Dusk Chorus* (2016) that follows David Monacchi's quest to record the diminishing sonic portraits of the world's ecosystems. This echoes Rachel Carson's *Silent Spring* (1962) where unnatural silence is the measure of environmental degradation.

Soundscape ecology is made possible by automated recording devices, inexpensive storage capabilities, and specialist software to analyse the complex recordings obtained. Soundscapes have most often been used to study bird communities, and have documented the decline of common bird species in Europe. They are now also used to evaluate the richness of insects, amphibians, mammals, and other vocalizing animals in natural habitats. Analytic challenges remain in distinguishing species from the cacophony of recorded sounds, but this exciting new coupling of research and technology has promise in delivering new perspectives on the diversity and dynamics of animal communities.

We are what we eat

Atoms of given elements often vary in the number of neutrons they contain, giving rise to different isotopes. 'Heavy' isotopes are neutron-rich, and 'light' if not. Unlike radioactive isotopes, non-radioactive stable isotopes do not decay. These stable isotopes are useful in ecology, as some isotopes are more readily assimilated by consumers, resulting in progressively changing isotope ratios in the bodies of animals up the food chain.

Isotope ratios, detected using mass spectrometry, allow inferences about the type and locality of food sources, and provide insights into the structures of marine and terrestrial food webs. Results are often unexpected. We now understand that omnivory, feeding on multiple trophic levels, is more prevalent than previously thought, challenging the view that organisms can be assigned to a particular trophic level. This has helped to resolve the 'ant-biomass paradox'. Ants are generally regarded to be predominantly predacious animals but, given the availability of animal food, seem disproportionately represented in insect samples from tropical tree crowns. Stable isotope studies suggest that canopy ants derive substantial amounts of their nitrogen requirements from herbivory rather than predation, by visiting nectar-secreting glands on plant leaves or stems, or by harvesting honeydew from aphids and other bugs. In Canadian lakes, isotopes implicated introduced invasive fish species, smallmouth bass and rock bass, in changes to the feeding behaviour of native trout from a primarily fish-based to a planktonic diet, demonstrating how invasive species restructure food webs to the detriment of native species. Stable isotopes have even uncovered how low-level nutrient inputs from tourists change lake food webs on Fraser Island, Australia, providing an impetus for park managers to improve toilet facilities!

Genetics

Ecologists spend more time in the laboratory than they might have done in the past. They have become adept at using genetic markers to track the movements and origins of seedlings, and mating patterns of plants and animals alike. This has proved crucial for understanding how changing land cover and forest clearance affects gene exchange among trees in a landscape, and hence the production of viable seed. Molecular techniques applied to DNA extracted from mammalian faeces can even provide information on population structure, food habits, reproduction, sex ratios, and parasite loads.

Advances in high-throughput sequencing allow simultaneous characterization of unique DNA sequences (or barcodes) from multiple species from environmental samples, with the diversity of genetic sequences providing indication of species numbers. This 'metabarcoding' technique has the potential to revolutionize assessments of the species composition within communities. Linking barcodes to particular species is currently limited, as reference libraries that associate DNA barcodes to species names are very incomplete. In some cases, taxonomy-free approaches are sufficient to gauge the diversity of poorly known assemblages such as marine plankton. Nonetheless, if ecology is to profit from DNA-based monitoring, metabarcoding needs to provide information on organismal traits and interactions, for which a taxonomic profile of detected DNA sequences will be necessary. Increasing the coverage of DNA reference libraries is simply a matter of time.

Remote sensing

Satellite imagery has been available since 1982 in the form of Landsat, with a 30 m resolution sufficient to detect broad land cover categories, and even distinguish large treefall gaps and storm blowdowns, important components of forest dynamics. Since the 1980s, satellite remote sensing has advanced considerably. New satellites provide regular coverage of the Earth's surface to resolutions as little as 30 cm. They detect not only vegetation cover, but also soil carbon, soil moisture, ocean salinity, and many other environmental variables of interest to ecology. Orbiting satellites track changes in vegetation cover, fire regimes, and animal movements over days to years.

Airborne sensors, while not having the same range as satellites, image vegetation cover at much higher resolution, and can even map plant traits and chemical diversity. Airborne LIDAR (Laser Imaging, Detection And Ranging) delivers three-dimensional imagery of vegetation structure and ground topography (digital elevation models) by detecting reflected pulsed laser light. Reduction in the size of the sensors now allows ground-based LIDAR to be deployed from a backpack.

Small, unmanned aerial vehicles (UAVs or drones) are now widely used for ecological research. They can map landscapes with centimetre resolution, and monitor vegetation cover and health in those landscapes. Data are collected more quickly, reliably, and cheaply over larger areas using UAVs than can ever be achieved by ground-based surveys. While they lack the geographic coverage of satellites, they do have much higher resolution and versatility, and can be flown close to areas of interest. This allows them to be used for plant species identification and wildlife surveys. Mammal population counts can be undertaken at discreet distances using UAVs controlled from the comfort of a car. The sleeping platforms of orang-utans in Borneo, and populations of elephants and rhinoceros in Africa, have been monitored in this way. UAVs are tracking and photographing whales, and analysing the images to understand how environmental conditions affect the health of adults, and the reproductive success of the population. While usually deployed singly, UAVs can be programmed to communicate with each other, allowing several to collect data simultaneously over larger areas.

Citizen science

For all their value, remote sensing technologies struggle to provide insights into the ecological processes or conditions under vegetation canopies or in the soil. For this, ecologists need ground-based sensors. Motivated people, coupled with smart phones, are excellent ground-based sensors. Technology has made it easy for anyone to collect ecological data, using little more than a mobile phone with integrated camera and GPS. Ecologists have welcomed 'citizen scientists' into a diverse array of research programmes which, through crowdsourcing, can leverage large data sets. Web-based platforms provide interfaces through which participants submit their data, perhaps simply by uploading a GPS-tagged photo. Citizen science is monitoring the spread of tree diseases and invasive alien species, thereby providing geographically extensive early warning systems. A variety of smartphone 'apps' are available for these purposes. Citizens are helping conservation agencies to map and expose illegal logging by uploading geo-referenced photos of such cases. Raising awareness of nature and its ecology through the lens of participatory research is a positive by-product of citizen science, and one that helps to build, and indeed rediscover, a culture of natural history in society.

Evolving ecology

Ecological science is evolving, in the questions it asks and the methods it uses. New technologies, from satellites to software, have increased the depth and breadth of ecological science, and provided new opportunities to engage with society. The prevalence of mathematical modelling and remotely sourced data absolves the ecologist from leaving the office chair for the rigours of a field campaign. The unfortunate consequence is a loss of connection to natural history—most ecologists would hardly know a sipunculid from a siphonophore. Yet our environment, and our relationship to it, needs to be understood in the context of the diverse array of interactions among plants, animals, and ecosystems. Without an intuitive sense of how nature works, gained through experiential natural history, we risk a loss of scientific creativity in ecology. The worry is that environmental management becomes informed predominantly by remotely sourced data that do not fully reflect the contextual richness of ecosystem interactions and contingencies. There is no substitute for the sedulous observation and diligent experimentation that characterizes ecological fieldwork.

An ecological fieldworker has been parodied, not altogether unfairly, as casually attired and slightly unkempt, the men hirsute (the women not), all seemingly comfortable in the company of creatures despised by others. Some ecologists turned television presenters have positively thrived on this caricature, endearingly turning it to their advantage. It remains undeniable that many, perhaps most, ecologists became ecologists through an early fascination with natural history and the outdoors. It is due to experiences so gained that ecologists become entwined with conservation. Aldo Leopold wrote, 'we can only be ethical in relation to something we can see, understand, feel, love, or otherwise have faith in'. It is thus essential to imbue children with ecologists' enthusiasm and curiosity for natural history. An appreciation of nature starts young, be it as an expression of delight at birds on the bird table, curiosity about moths flittering around a light, or excitement with tadpoles in a pond (Figure 31). Such interests might be fleeting, but early exposure to the

31. An early interest in tadpoles can foster an environmental awareness that this planet so desperately needs.

variety of life, and to natural history, fosters a keener sense of environmental awareness and responsibility. An appreciation of natural history gained early is longer lasting and more meaningful, and it is to the education of our children that we must devote attention and energy if we are to achieve a truly sustainable society. It is from the clay of natural history that ecologists are moulded.

Further reading

There is no end to the possibilities for further enquiry into the broad field of ecology across the range of its interpretations and applications. My suggestions loosely reflect the themes covered in this short book, which are clearly far from comprehensive. I have selected a number of books and articles that fall into three main categories: those that are classical markers in the development of the ecological discipline and that continue to have a lasting legacy; others that are accessible and informative treatments of general concepts; and more mainstream textbooks as might be adopted by standard university ecology courses.

Chapter 1: What is ecology?

Beeby, A. and Brennan, A. M. (2004) *First Ecology: Ecological Principles and Environmental Issues*. 2nd edition. Oxford University Press, Oxford. 318 pages.
A good introductory text to ecological science.

Begon, M., Townsend, C. R., and Harper, J. L. (2005) *Ecology: From Individuals to Ecosystems*. 4th edition. Wiley-Blackwell, Hoboken, NJ. 750 pages.
A long-standing textbook for undergraduate ecological courses, and rather more comprehensive than Beeby & Brennan. Not exactly entertaining reading, but highly informative, especially on ecology as a theoretical and experimental science.

Hagen, J. B. (1992) *An Entangled Bank: The Origins of Ecosystems Ecology*. Rutgers University Press, New Brunswick, NJ. 245 pages.

Accessible and well-written book on ecological theory, as traced through its historical development since 1900, albeit with a North American bias.

Scheiner, S. M. and Willig, M. R. (eds) (2011) *The Theory of Ecology*. University of Chicago Press, Chicago. 416 pages.

The science of ecology has a strong theoretical basis, although this can be difficult to discern. *The Theory of Ecology* is a series of contributions that aims to convey theoretical clarity and structure to ecological theory.

Chapter 2: The dawn of ecology

Anderson, J. G. T. (2013) *Deep Things out of Darkness: A History of Natural History*. University of California Press, Berkeley.

Evaluates the role of natural history in the development of ecological science and environmental discourse, and argues for the need for natural history in our current era of environmental change.

Clements, F. E. (1936) Nature and structure of the climax. *Journal of Ecology* 24: 252–84.

Frederic Clements's perspective on succession as a developmental process whose final stage, the climax formation, is determined primarily by regional climate, with all other types of vegetation formations constituting stages of development on the path towards the climax state. While this view is no longer accepted, the concept of a successional process remains highly relevant and influential.

Connell, J. H. (1961) The influence of interspecific competition and other factors on the distribution of the barnacle *Chthamalus stellatus*. *Ecology* 42: 710–23.

A seminal study on how competition among two species, coupled with their different proclivities to local physical conditions, can shape their distributions.

Elton, C. (1927) *Animal Ecology*. Macmillan Press, New York.

An early classic text in which Elton defined some of the foundational ecological concepts, including that of an ecological community interpreted through the trophic interactions that occur between its living components.

Hutchinson, G. E. (1957) Concluding remarks. *Cold Spring Harbour Symposium on Quantitative Biology* 22: 415–27.

The paper that formalized the definition of an ecological niche.

Hutchinson, G. E. (1959) Homage to Santa Rosalia, or why are there so many kinds of animals? *The American Naturalist* 93: 145–59.
    Addresses the issue of how species avoid competitive exclusion to coexist within a seemingly similar environment.

Kricher, J. (2009) *The Balance of Nature: Ecology's Enduring Myth.* Princeton University Press, Princeton. 256 pages.
    Very readable exploration of the history of ecology with particular emphasis on the debate regarding concepts of self-regulation in ecological systems.

Worster, D. (1994) *Nature's Economy: A History of Ecological Ideas.* 2nd edition. Cambridge University Press, Cambridge. 526 pages.
    An excellent history of the field of ecosystem ecology.

Chapter 3: Populations

Hanski, I. (1999) *Metapopulation Ecology.* Oxford University Press, Oxford.
    A thorough synthesis of research on metapopulations, drawing on the author's own research on the Glanville fritillary butterfly. Includes discussion of the relevance of metapopulation ideas to conservation biology.

Rockwood, L. L. (2015) *Introduction to Population Ecology,* 2nd edition. Wiley-Blackwell, Hoboken, N.J. 378 pages.
    A textbook treatment of population ecology, drawing on a wide array of examples and experiments to explore the fundamental laws of population ecology, including the role of interactions such as competition, mutualism, predation, and herbivory.

Vandermeer, J. H. and Goldberg, D. E. (2013) *Population Ecology: First Principles.* Princeton University Press, Princeton. 263 pages.
    A quantitative approach that presents some of the mathematical and theoretical foundations underlying the structure and dynamics of populations.

Chapter 4: Communities

Bronstein, J. L. (ed.) (2015) *Mutualism.* Oxford University Press, Oxford. 320 pages.
    An authoritative perspective on the ecology and evolution of mutualisms.

Eichhorn, M. P. (2016) *Natural Systems: The Organisation of Life.* Wiley Blackwell, Hoboken, NJ. 392 pages.

A textbook treatment that encompasses the links between ecology, biodiversity, and biogeography.

Estes, J. A. (2016) *Serendipity: An Ecologist's Quest to Understand Nature.* University of California Press, Berkeley. 256 pages.

James Estes's personal narrative on a fifty-year research career on trophic cascade ecology, based on his field studies on the Aleutian Islands examining relationships among kelp forests, sea otters, sea urchins, and killer whales.

Pimm, S. L. (1991) *The Balance of Nature? Ecological Issues in the Conservation of Species and Communities.* The University of Chicago Press, Chicago. 448 pages.

Stuart Pimm provides an ecological critique and analysis of widely used terms such as 'stability', 'balance of nature', and 'resilience', and places the interpretation of these terms in the context of the food web structures and the physical environment.

Silvertown, J. (2005) *Demons in Eden: The Paradox of Plant Diversity.* University of Chicago Press, Chicago. 169 pages.

In this very readable book, Silvertown links ecological and evolutionary processes to understand the emergence and maintenance of plant diversity through the interacting effects of environmental conditions, species competition, predation, and dispersal.

Chapter 5: Simple complex questions

Colinvaux, P. (1978) *Why Big Fierce Animals are Rare: An Ecologist's Perspective.* Princeton University Press, Princeton. 256 pages.

An outstanding collection of essays that delve into many ecological and biological ideas, of which *Why Big Fierce Animals Are Rare* is but one. The book covers many big ideas in ecology, including ecosystems, habitats, communities, niches, associations, and animal population dynamics.

Connell, J. H. (1971) On the role of natural enemies in preventing competitive exclusion in some marine animals and in rain forest trees. In: P. J. Den Boer and G. R. Gradwell (eds), *Dynamics of Population.* Pudoc, Wageningen.

Janzen, D. H. (1970) Herbivores and the number of tree species in tropical forests. *The American Naturalist* 104: 501–28.

Dan Janzen and Joseph Connell independently formulated the idea that density-dependent processes mediated by natural enemies, such as seed predators, could maintain species coexistence. The subsequently named Janzen–Connell model continues to be highly influential and a widely accepted mechanism explaining high species diversity in the tropics.

Sherratt, T. N. and Wilkinson, D. M. (2009) *Big Questions in Ecology and Evolution*. Oxford University Press, Oxford. 312 pages.
Following the spirit of Colinvaux's book *Why Big Fierce Animals are Rare*, Sherratt and Wilkinson's book considers a range of fundamental ecological and evolutionary questions that continue to be discussed and debated.

Chapter 6: Applied ecology

Baskin, Y. (2003) *A Plague of Rats and Rubbervines: The Growing Threat of Species Invasions*. Island Press, London. 330 pages.
An engaging exploration of invasive alien species and the problems they have caused across the world, and the efforts invested in trying to control them.

Gunderson, L. H., Allen, C. R., and Holling, C. S. (eds) (2012) *Foundations of Ecological Resilience*. Island Press, Washington, DC. 496 pages.
Ecological resilience theory provides a basis for understanding how complex systems adapt to and recover from disturbances. *Foundations of Ecological Resilience* is a collection of some of the most influential articles on ecological resilience.

Newman, E. I. (2001) *Applied Ecology and Environmental Management*. 2nd edition. Wiley-Blackwell, Hoboken, NJ. 408 pages.

Townsend, C. R. (2007) *Ecological Applications: Toward a Sustainable World*. Wiley-Blackwell, Hoboken, NJ. 328 pages.
Two books that describe and present issues in the realm of environmental management and sustainability, drawing on ecological theory at individual, populations, and community levels

Wilson, E. O. (1988) *Biodiversity*. Harvard University Press, Cambridge, Mass.
An edited volume with contributions on the current threats to biodiversity, its value, and practical and policy approaches to conserving and restoring biodiversity.

Chapter 7: Ecology in culture

Berkes, F. (2008) *Sacred Ecology*. 2nd edition. Routledge, Abingdon. 313 pages.

Describes and evaluates the contributions of traditional ecological knowledge to natural resource management. Reflects growing interest in alternative ecological visions and insights from indigenous resource use practices, and the need to develop a new ecological ethic by drawing on a wide range of traditions.

Carson, R. (1962) *Silent Spring*. Houghton Mifflin Co., Boston. 378 pages.

'Every once in a while in the history of mankind, a book has appeared which has substantially altered the course of history': Senator Ernest Gruening of Alaska. He was referring to *Silent Spring*.

Leopold, A. (1949) *A Sand County Almanac: And Sketches Here and There*. Oxford University Press, Oxford. 226 pages.

Little more than a series of personal thoughts and reflections of nature and our interactions with it, yet immensely powerful and influential, contributing greatly to the development of modern conservation science, policy, and ethics.

Lovelock, J. (1979) *Gaia: A New Look at Life on Earth*. Oxford University Press, Oxford. 148 pages.

A much debated but classic work that continues to inspire many and aggravate some. James Lovelock argues that life on earth functions as if it were a single self-organizing organism.

Naess, A. (1989) *Ecology, Community, and Lifestyle: Outline of an Ecosophy*. Translated and edited by D. Rothenberg. Cambridge University Press, Cambridge. 223 pages.

Naess argues that environmental issues are framed by the values of people and society, and that such values are shaped by ethical considerations. He advocates that we should conceive ourselves as part of the world whereby the value of life and nature is intrinsic to our being, an approach based on 'deep ecological principles'.

Chapter 8: Future ecology

Dayton, P. K. (2003) The importance of the natural sciences to conservation. *American Naturalist* 162: 1–13.

A comprehensive argument calling to reinstate basic knowledge of natural history in science courses to enable meaningful understanding of, and action for, environmental management and conservation.

Devictor, V., van Swaay, C., Brereton, T., Brotons, L., Chamberlain, D., Heliölä, S., Herrando, J., Julliard, R., Kuussaari, M., Lindström, Å., Reif, J., Roy, D. B., Schweiger, O., Settele, J., Stefanescu, C., Van Strien, A., Van Turnhout, C., Vermouzek, Z., WallisDeVries, M., Wynhoff, I., and Jiguet F. (2012) Differences in the climatic debts of birds and butterflies at a continental scale. *Nature Climate Change* 2: 121–4.

A paper that compared the rates at which bird and butterfly communities could keep up with temperature change across Europe. Concludes that both birds and butterflies are not able to keep up with temperature increases, implying the accumulation of 'climatic debts' for these groups at continental scales.

Hampton, S. E., Strasser, C. A, Tewksbury, J. J., Gram, W. K., Budden, A. E., Batcheller, A. L., Duke, C. S., and Porter, J. H. (2013) Big data and the future of ecology. *Frontiers in Ecology and the Environment* 11: 156–62.

Ecologists are generating increasingly large volumes of data by a variety of means, but there is little collective planning on how such data are curated. This article advocates that if ecologists are to address large-scale complex questions of the future, they will need to organize and archive data for posterity, share their data freely, and participate in collaborations among scientists and the wider public.

Other texts on topics that this Ecology VSI has not been able to address in detail include:

Brown, J. H. (1995) *Macroecology*. Chicago University Press, Chicago. 269 pages.

Ecological processes give rise to patterns in nature, but mostly these have been explored at relatively small spatial scales that are amenable to observation and experimentation. The field of macroecology extends the discipline to much larger spatial and temporal scales, to explore patterns of life across the globe.

Crawley, M. J. (ed.) (1997) *Plant Ecology*. Blackwell Science, Oxford. 736 pages.

An excellent edited collection of contributions that encompasses broad themes in plant ecology, including ecophysiology, population dynamics, community structure, ecosystem function, herbivory, sex, dispersal, global warming, pollution, and biodiversity.

Ghazoul, J. and Sheil, D. (2010) *Tropical Rain Forest Ecology, Diversity, and Conservation*. Oxford University Press, Oxford. 516 pages.
An introduction to the tropical rainforests of the world, their species diversity, and the richness of ecological interactions that sustain them.

Whittaker, R. J. (1998) *Island Biogeography: Ecology, Evolution, and Conservation*. Oxford University Press, Oxford. 285 pages.
Biogeography has its roots in ecology. Island biogeography theory was originally developed by Robert MacArthur and E. O. Wilson as the *Theory of Island Biogeography* (1967) to explain the species richness and dynamics of islands. This topic has become very influential in conservation theory, and especially in the debate on the size and number of protected areas.

With, K. A. (2019) *Essentials of Landscape Ecology*. Oxford University Press, Oxford. 656 pages.
Landscape ecology is the science that investigates ecological processes and patterns in natural and human structured landscapes across a wide range of scales.

Index

For the benefit of digital users, indexed terms that span two pages
(e.g., 52–53) may, on occasion, appear on only one of those pages.

Index

GEOGRAPHY
A Very Short Introduction
John A. Matthews & David T. Herbert

Modern Geography has come a long way from its historical roots in exploring foreign lands, and simply mapping and naming the regions of the world. Spanning both physical and human Geography, the discipline today is unique as a subject which can bridge the divide between the sciences and the humanities, and between the environment and our society. Using wide-ranging examples from global warming and oil, to urbanization and ethnicity, this *Very Short Introduction* paints a broad picture of the current state of Geography, its subject matter, concepts and methods, and its strengths and controversies. The book's conclusion is no less than a manifesto for Geography' future.

> 'Matthews and Herbert's book is written- as befits the VSI series- in an accessible prose style and is peppered with attractive and understandable images, graphs and tables.'
>
> **Geographical.**

ONLINE CATALOGUE
A Very Short Introduction

Our online catalogue is designed to make it easy to find your ideal Very Short Introduction. View the entire collection by subject area, watch author videos, read sample chapters, and download reading guides.

http://fds.oup.com/www.oup.co.uk/general/vsi/index.html

SOCIAL MEDIA
Very Short Introduction

Join our community
www.oup.com/vsi

- Join us online at the official Very Short Introductions
  Facebook page.
- Access the thoughts and musings of our authors with our
  online **blog**.
- Sign up for our monthly **e-newsletter** to receive information
  on all new titles publishing that month.
- Browse the full range of Very Short Introductions online.
- Read **extracts** from the Introductions for free.
- Visit our library of **Reading Guides**. These guides, written by our
  expert authors will help you to question again, why you think
  what you think.
- If you are a teacher or lecturer you can order inspection
  copies quickly and simply via our website.